Maneuvers with Circles
Student Lab Book

David A. Page
Kathryn Chval

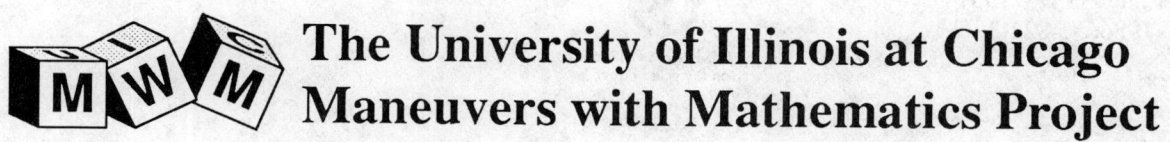

**The University of Illinois at Chicago
Maneuvers with Mathematics Project**

DALE SEYMOUR PUBLICATIONS

Other UIC-MWM Student Lab Books
Maneuvers with Rectangles
Maneuvers with Angles
Maneuvers with Triangles
Maneuvers with Nickels and Numbers

(A Teacher Sourcebook is also available for each Student Lab Book in the series.)

The Maneuvers with Mathematics Project materials were prepared with the support of National Science Foundation Grant Nos. MDR-8850466 and MDR-9154110. Any opinions, findings, conclusions, or recommendations expressed in this publication are those of the authors and do not necessarily represent the views of the National Science Foundation. These materials shall be subject to a royalty-free, irrevocable, worldwide, nonexclusive license in the United States Government to reproduce, perform, translate, and otherwise use and to authorize others to use such materials for Government purposes.

Copyright © 1995, David A. Page. All rights reserved. Portions of this work were published as field trial editions in 1992, 1993, copyright David A. Page. No part of this publication may be reproduced, stored in a retrieval system, or transmitted, in any form or by any means, electronic, mechanical, photocopying, recording, or otherwise, without prior written permission of the authors. Printed in the United States of America.

Order number DS21331
ISBN 0-86651-822-3

4 5 6 7 8 9 10-MA-98 97 96

This Book Is Printed on Recycled Paper

Contents

1. Circumnavigate a Circle 1
2. What's Inside? 17
3. In and About 29
4. Pieces of Circles 45
5. Remember That Number 55
6. Circles, Sectors, and Angles 69
7. Circles with Holes 89
8. Perimeter of Pieces 109
9. Circles, Chords, and the Pythagorean Theorem 123
10. Grazing Goats 135

Essential contributions to UIC-MWM were made by:

John Baldwin
Roberta Dees
Steven Jordan

Janice Banasiak
Marty Gartzman
Olga Granat-Gonzalez
Michael Jankowski
Jennifer Lynná Mundt
Marlynne Nishimura
Pamela Piggeé
Jerome Pohlen
Mary Jo Porn
Aimee W. Strawn
Mary Ann Schultz

Production Assistants:

Lindy M. Chambers
Kimberly Hanus
Tanya Henderson
Tracy Ho
Alex Mak
Stacie McCloud
Erik Merkau
Olga Vega
Wendy Wisneski

Artists:

Lisa Fucarino
Alex Mak

The University of Illinois-Maneuvers with Mathematics (UIC-MWM) project started in July 1989 under the direction of David A. Page and Philip Wagreich of UIC. Earlier versions were tested in the following schools in Illinois:

Louis J. Agassiz Elementary School, Chicago
Albright Middle School, Villa Park
Caroline Bentley School, New Lenox
Daniel Boone Elementary School, Chicago
Carpenter School, Park Ridge
Central Jr. High School, Tinley Park
Christ the King School, Lombard
Walt Disney Magnet School, Chicago
Richard Edwards Elementary School, Chicago
Eugene Field Elementary School, Park Ridge
John Hope Community Academy, Chicago
Andrew Jackson Language Academy, Chicago
John L. Marsh Elementary School, Chicago
Our Lady of Victory School, Chicago
John Palmer School, Chicago
W. C. Petty School, Antioch
Pilsen Community Academy, Chicago
Philip Rogers School, Chicago
St. Germaine School, Oak Lawn
St. Joseph School, Chicago
St. Michael the Archangel School, Chicago
St. Stephen Protomartyr School, Des Plaines
Mark Sheridan Math & Science Academy, Chicago
Wendell Smith Elementary School, Chicago
John M. Smyth Elementary School, Chicago
Washington Elementary School, Park Ridge

Name _____ Date _____ Class _____   1

1. Circumnavigate a Circle

1. Place a check below each *circle* in the following figure.

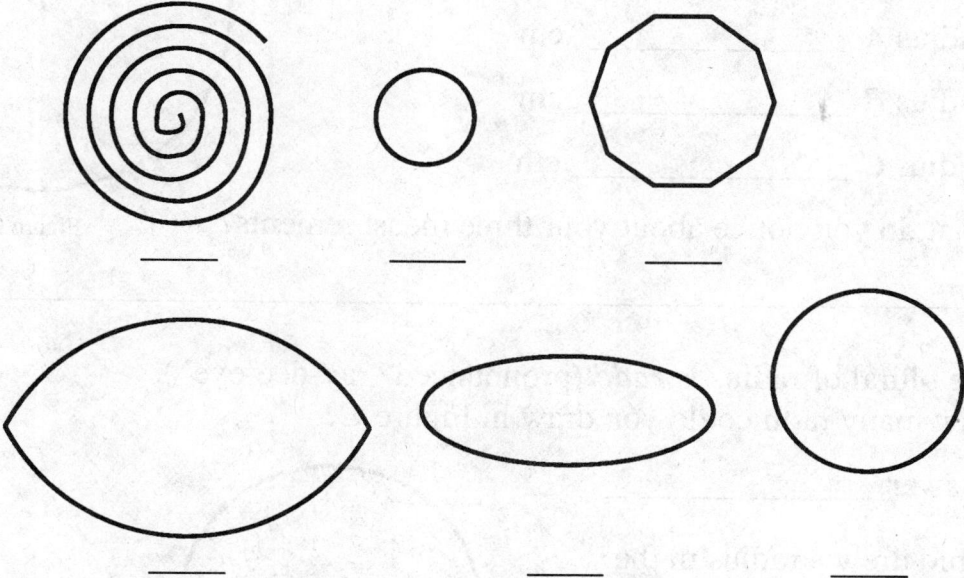

Figure A

2. How would you describe a circle? Make your description clear so that the other figures do not qualify.

_____

_____

 3. The distance from the **center** of a circle to its **rim** is called a **radius**. Measure each radius to the nearest 0.1 cm in the following figure.

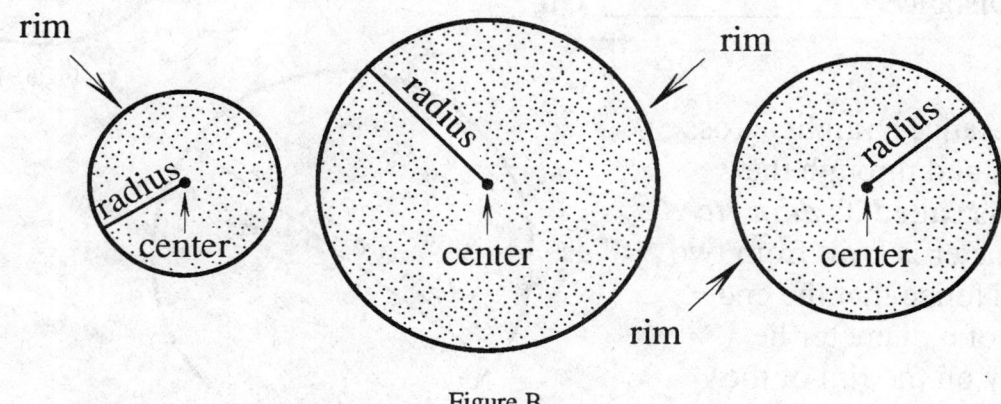

Figure B

_____ cm   _____ cm   _____ cm

© David A. Page   Maneuvers with Circles

4. A radius is drawn in the circle at the right. Draw another radius and label it "Radius B." Draw a third radius and label it "Radius C." Measure each radius to the nearest 0.1 cm.

4a. Radius A _____ cm

4b. Radius B _____ cm

4c. Radius C _____ cm

4d. What do you notice about your three measurements?
_____

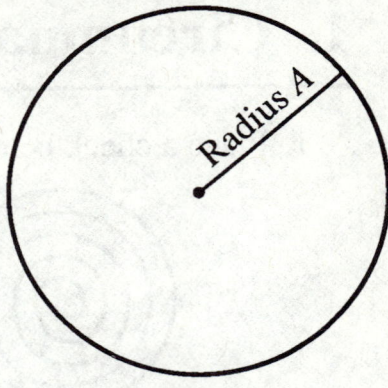

Figure C

5. The plural of radius is *radii* (pronounced "ray-dee-eye"). How many radii could you draw in Figure C?

Answer _____

6a. Pablo drew a radius in the circle in Figure D. Measure Pablo's radius to the nearest 0.1 cm.

Radius _____ cm

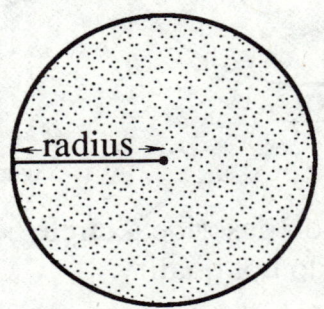

Figure D

6b. Pablo then drew a second radius next to the first radius. Without measuring, what is the distance straight across his circle?

Distance _____ cm

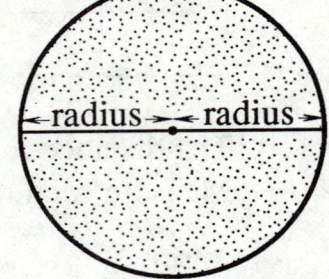

Figure E

The distance straight across a circle and through the center is called its *diameter*. A diameter is built from two radii. Notice that the end points of a diameter lie directly on the rim of the circle.

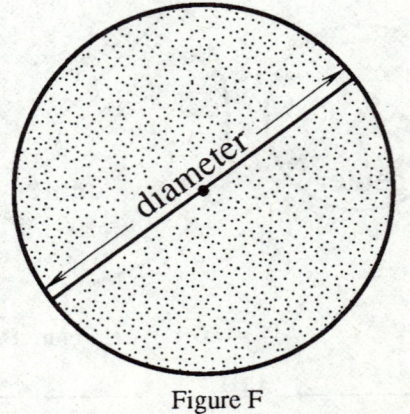

Figure F

Maneuvers with Circles © David A. Page

Circumnavigate a Circle 3

7a. Place a check mark next to each circle which has a diameter drawn in it.

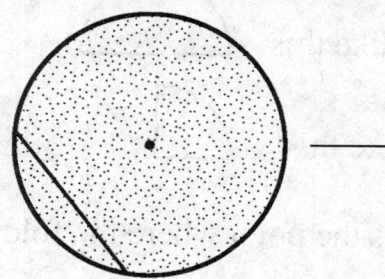 ____ 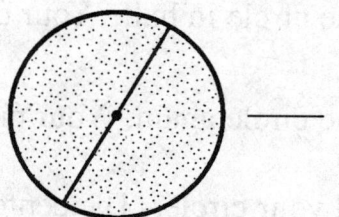 ____

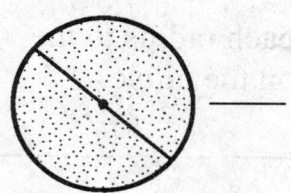 ____ 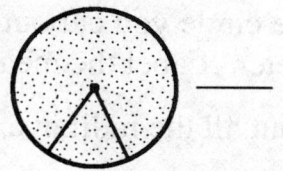 ____ 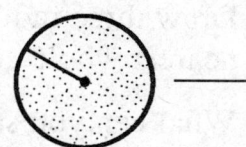 ____

Figure G

7b. Look at the circles you did not check above. Underneath those circles, explain why the line or lines are not diameters.

8a. How many diameters are drawn in the circle at the right?

Answer _____

8b. How many radii are drawn?

Answer _____

Figure H

 8c. Measure the diameter to the nearest 0.1 cm.

Diameter _____
Put in units.

8d. Measure the radius to the nearest 0.1 cm.

Radius _____ cm

8e. How would you now describe a circle?

_____

_____

© *David A. Page*   *Maneuvers with Circles*

9a. Cut out the circle at the bottom of the page.

9b. Fold the circle in half. Your circle will look like this:

9c. Fold the circle again. Your circle will look like this:

9d. Unfold your circle. The center of the circle is the point where the folds meet. Draw a dot there.

 10a. Draw three radii in the circle you cut out. Measure each radius to the nearest whole centimeter. Label the measurements on the circle.

10b. What can you say about all the radii in a circle? _____

11a. Draw three diameters and measure each one.

11b. What can you say about all the diameters in a circle? _____
_____

12. Can you draw a straight line in the circle that is longer than the diameter?

Answer _____
_{Yes or No}

Why or why not? _____

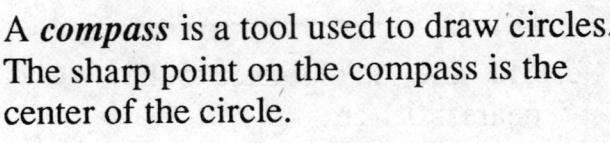

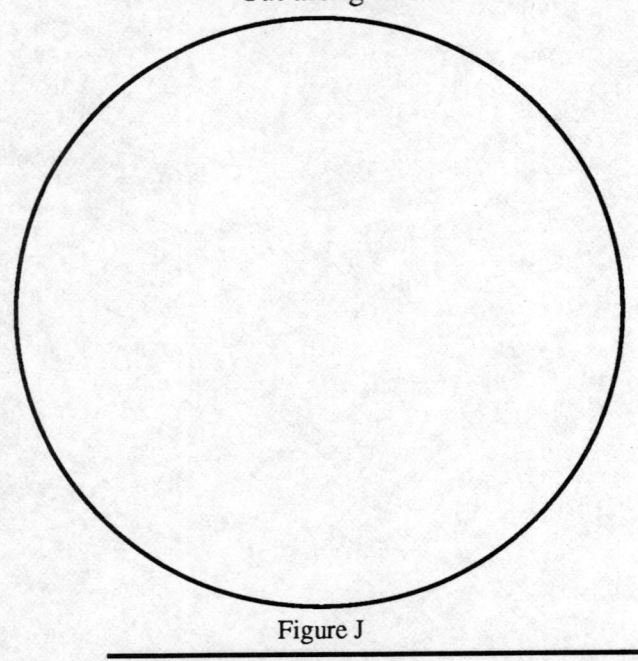

Figure J

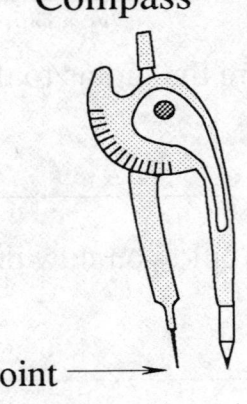

Compass

point

Maneuvers with Circles © *David A. Page*

Circumnavigate a Circle 5

 13. Draw a circle below using the following steps.

    a. Place the sharp point of the compass in the middle of the —|—. This will be the center of the circle.

    b. Keep the point on the center and open the compass until the pencil touches the center of the dot (•).

    c. Without lifting the point, turn the compass to draw the circle.

14. Draw a diameter in your circle. Measure it to the nearest 0.1 cm.

    Answer _____ cm

15. Without measuring, how long is the radius? _____ cm

16a. Draw a different circle using the same center above, by opening or closing the compass.

16b. Measure the new diameter. _____ cm

17a. Draw a third circle using the same center above.

17b. Measure the radius of your third circle. _____ cm

© David A. Page   *Maneuvers with Circles*

18. Use the steps and sketches in the following table along with the center mark below to draw a circle with a radius of 4 cm.

		Steps	Sketches
	a.	Use your ruler to draw a dot 4 cm away from the center mark.	
	b.	Place the sharp point of the compass on the center mark. Then open the compass so that the sharp point is on the center and the pencil is on the dot you drew.	
	c.	Without lifting the point, turn the compass to draw the circle.	

19a. Draw a radius in the circle you drew below.

19b. Measure the radius to check.

First radius _____ cm

19c. Draw and measure a different radius.

Second radius _____ cm

19d. How long is the diameter?

Diameter _____ cm

19e. Draw and measure a diameter.

Diameter _____ cm

Circumnavigate a Circle

20a. Draw a circle with a radius of 2.4 cm.

20b. Draw a diameter.

20c. Without measuring, what is the length of the diameter?

Diameter _____
Put in units.

20d. How did you find the length of the diameter?

_____

_____

21a. Draw a circle with a diameter of 5.6 cm. Measure the diameter to check.

21b. Draw a radius.

21c. Without measuring, what is the length of the radius?

Radius _____ cm

21d. How did you find the length of the radius? _____

_____

22. Figure K is a sketch. What is its radius?

Radius _____ cm

23. A different circle has a radius of 617,283.5 cm. What is the diameter of this circle?

Diameter _____ cm
You'll know.

Figure K (26.55 cm)

24. Jenna's pool has a diameter of 114 cm. What is the radius of her pool?

Radius _____ cm

© David A. Page

Maneuvers with Circles

Perimeter and Circumference

A *polygon* is a closed shape which has three or more straight sides. Figure L shows three examples of polygons.

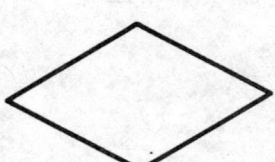 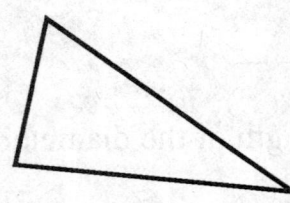 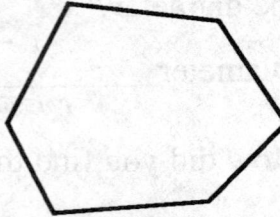

Figure L

1. Measure each side of the polygons in Figure L to the nearest 0.1 cm and label them on the figure.

The following shapes are special polygons called *regular polygons*.

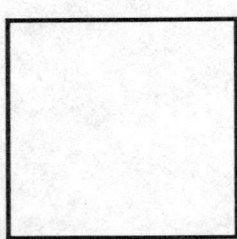 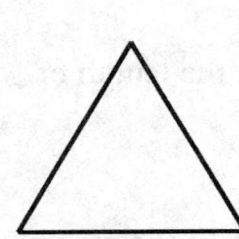 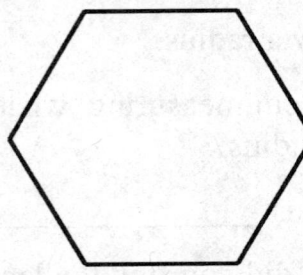

Figure M

2. Measure each side of the regular polygons in Figure M to the nearest 0.1 cm and label them on the figure.

3. What is special about a regular polygon? _____
_____

The distance around the outside of a shape is called the *perimeter*. Imagine a bug crawling around the outside of the polygon at the right. The perimeter is the distance the bug crawls to get all the way back to the point where it started.

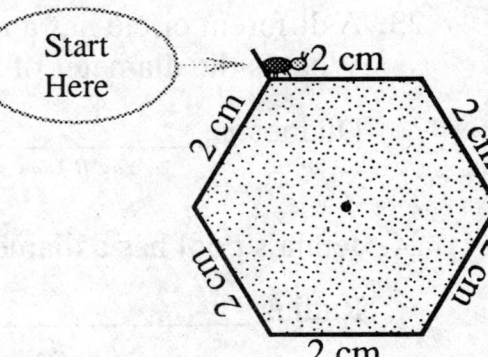

4. What is the perimeter of this polygon?

   Perimeter _____ cm

Figure N

5. Find the perimeter of each polygon in Figure L and Figure M. Write the perimeter inside each polygon.

Maneuvers with Circles © *David A. Page*

Circumnavigate a Circle

Unlike polygons, the distance around a circle is called its *circumference*. The circumference is the distance the bug crawls to get all the way back to the point where it started.

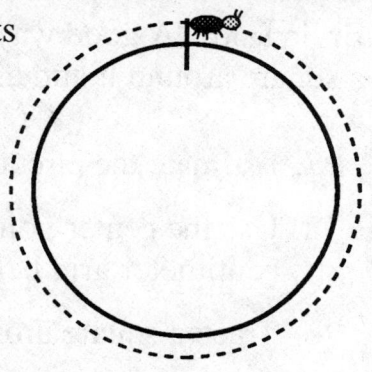

Figure P

If you draw a circle on the sidewalk with chalk and walk around it, you are walking on the circle's circumference.

Alfredo drew the following circle with his compass. The distance his pencil traveled to form the circle is the circumference.

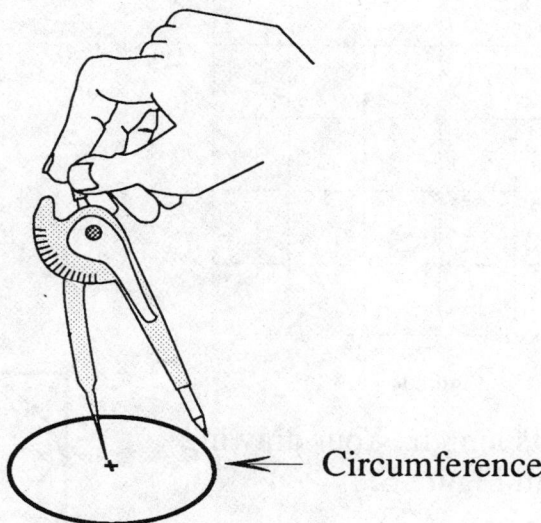

← Circumference

6. Michelle and Sandor wrapped a piece of string around the circle in Figure Q and then measured the string to find its circumference. What other way could you find the circumference of the circle at the right?

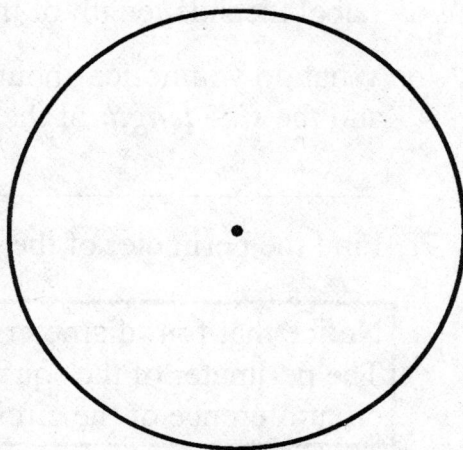

Figure Q

_____

_____

_____

_____

© David A. Page

Maneuvers with Circles

10 Chapter 1

Ervin said, "A good way to estimate the circumference of a circle is to build a square around it and use the square's perimeter as an estimate."

7. Estimate the circumference of the circle using the following steps.

7a. Use the center point to draw a circle with a radius of 3 cm on the centimeter grid below.

7b. Trace a square around the circle.

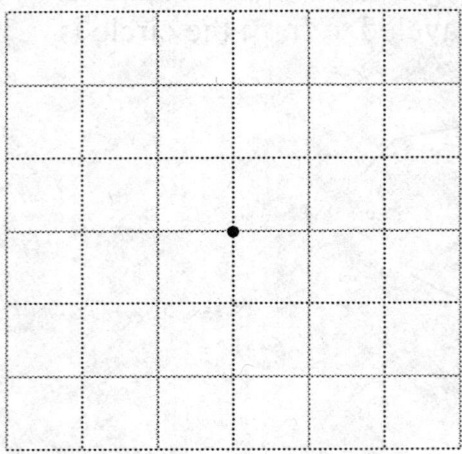

Figure R

7c. Draw a diameter and label its length. Your drawing should look like the sketch in Figure S.

7d. Label the side length of the square.

7e. What do you notice about the length of the *diameter* and the *side length* of the square?

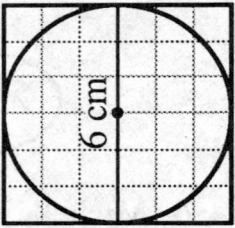

Figure S

_____

7f. Find the perimeter of the square. _____ cm

> Notice that four diameters of the circle form the perimeter of the square. The perimeter of the square built around a circle is one estimate for the circumference of the circle.

8. Is the circle's circumference smaller or larger than the perimeter of the square?

   Answer _____
   Smaller or Larger

Maneuvers with Circles © *David A. Page*

Circumnavigate a Circle 11

Lance said, "To find another estimate for the circumference of a circle, draw a ***regular hexagon*** inside of it." A regular hexagon is a polygon with six equal sides.

9. Use the following steps to draw a regular hexagon inside the circle on page 10.

a. Set your compass for the length of the radius, 3 cm. Once you set that distance, make sure the compass doesn't change.

b. Put your compass point on the rim of the circle and make a starting mark as shown at the right above.

c. Place your compass point on the starting mark and make a new mark along the rim.

d. Continue this until you get back to where you started.

e. Place a dot where each mark touches the rim.

f. Connect the dots with straight lines to form a regular hexagon like the sketch in Figure T. Label the length of one side.

Figure T

10. Find the perimeter of the regular hexagon. _____ cm

> Notice that 6 radii, which is also the same as 3 diameters, build the perimeter of the regular hexagon. The perimeter of the hexagon is another estimate for the circumference of the circle.

11. Is the circle's circumference smaller or larger than the perimeter of the regular hexagon?

　　Answer _____
　　　　Smaller or Larger

The circumference of a circle is smaller than the square's perimeter and larger than the hexagon's perimeter. Consider the circle with a radius of 3 cm.

12. The circumference of the circle is between _____ and _____.

13. Estimate the circumference of the circle. _____ cm

14. Measure the circumference of the circle. _____ cm

© David A. Page　　　　　　　　　　　　　　　　　　　　　　*Maneuvers with Circles*

15. Estimate the circumference of the circle at the right using the following steps.

15a. Measure the perimeter of the square around the circle.

   Answer _____ cm

15b. Measure the perimeter of the regular hexagon inside the circle.

   Answer _____ cm

15c. Estimate the circumference of the circle.

   Answer _____ cm

Figure U

15d. Elijah said, "I didn't use the six radii to estimate the circumference of the circle. I used the diameter of the circle instead."

   How did Elijah estimate the circumference of the circle? _____

   _____

Complete the following steps to collect data that will show a relationship between the diameter of a circle and its circumference.

16a. As a class, gather five different cylinders. For example, you can use a coffee can, an oatmeal carton, and a frozen orange juice can. Label the cylinders "1", "2", "3", "4", and "5". Divide into teams. Each team will work with one cylinder.

16b. Draw the diameter on the circle of each cylinder. Be sure you run across the center to get a diameter and not a shorter line.

17. If the center of the circle is not marked, how do you draw a diameter?

   _____

   _____

Maneuvers with Circles © David A. Page

Circumnavigate a Circle   13

 18. Measure the diameter of your team's cylinder to the nearest 0.1 cm.

Diameter _____ cm

19. Use a measuring tape to find the circumference of your team's cylinder.

Circumference _____ cm

20a. Collect the data from your classmates and complete the following table.

Cylinder	Measured Diameter *Put in units.*	Measured Circumference *Put in units.*
1		
2		
3		
4		
5		

20b. Compare the diameter and circumference columns. For each cylinder, about how many diameters equal the circumference?

Answer _____

21. Liz measured the diameter on her cylinder. Then she cut pieces of string that equaled the diameter. Liz taped the pieces of string along the circumference as shown in the following picture.

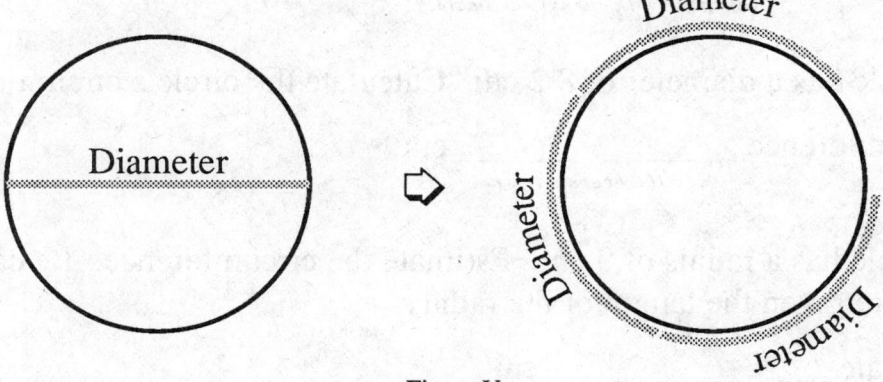

Figure V

21a. Use string and Liz's method on one of your cylinders.

21b. Is the circumference built from three diameters? _____
    Yes or No

Explain. _____

© David A. Page   *Maneuvers with Circles*

14   Chapter 1

> Using measurements, Circumference ÷ Diameter is a little more than 3. To do careful work with circles, you must work with the number π, which is spelled *pi* (pronounced "pie"). Press $\boxed{\pi}$ on your calculator.
>
> Window: $\boxed{\boxed{}.\boxed{}\boxed{}\boxed{}\boxed{}\boxed{}\boxed{}}$
>
> Notice that π is a little more than 3.
>
> You do not need to memorize π because your calculator knows π to 7 decimal places.
>
> For every circle: Circumference ÷ Diameter = π
>
> To *estimate* the circumference of a circle, multiply the diameter by 3.
> To *calculate* the circumference of a circle, multiply the diameter by π.

22a. Jenny drew a circle with a diameter of 5 cm. To estimate the circumference of the circle, she multiplied 5 cm by 3.

   Jenny's estimate _____ cm

22b. To calculate the circumference of her circle, Jenny pressed:

   $\boxed{5}$ $\boxed{\times}$ $\boxed{\pi}$ $\boxed{=}$

   Circumference _____ cm
   _{Copy window.}
   _{Compare with Jenny's estimate.}

23. A circle has a diameter of 8.2 cm. Calculate the circle's circumference.

   Circumference _____ cm
   _{R to the nearest tenth.}

24a. A circle has a radius of 3 cm. Estimate the circumference. Be careful! You are given the length of the radius.

   Estimate _____ cm

24b. Calculate the circumference.

   Circumference $\boxed{\boxed{}\boxed{}.\boxed{}\boxed{4}\boxed{9}\boxed{}\boxed{}}$ cm

Maneuvers with Circles   © *David A. Page*

Name _____ Date _____ Class _____   15

Homework 1: Circumnavigate a Circle

 1a. Draw a circle with a radius of 3.1 cm using the center at the right.

1b. Measure the circumference to the nearest 0.1 cm.

Circumference _____ cm

 2a. Draw a circle with a diameter of 7.8 cm below.

2b. Measure the circumference to the nearest 0.1 cm.

Circumference _____ cm

 3. A compact disc has a radius of 6 cm. Circle the best estimate for the circumference of the compact disc.

6 cm   12 cm   18 cm   24 cm   36 cm   50 cm

4. Calculate the circumference of the compact disc.

Circumference _____ cm
Copy window.

5. A circle has a diameter of 12.2 cm. Calculate its circumference.

Circumference _____ cm
Copy window.

© David A. Page   *Maneuvers with Circles*

6. Complete the following table. Remember to include the units. List the keystrokes you used to find the circumference.

	Radius	Diameter	Circumference *Copy window.*	Keystrokes
a.	2.75 cm			
b.		11 cm		
c.	11 cm			
d.		44 cm		
e.	44 cm			

7a. In the table above, what happens to the length of the radius?

_____

7b. What happens to the length of the diameter?

_____

7c. What happens to the circumference?

_____

8a. The Ridge Street Block Club is planting a circular garden in an empty lot. The diameter of the garden will be 40 ft. A fence around the garden will protect it from visiting creatures (rabbits) during the night. How many feet of fence is needed?

Length ☐ ☐ ☐ . 6 6 ☐ ☐ ☐ ft

8b. The store only sells fence by the whole foot. How many whole feet should the Ridge Street Block Club buy?

Answer _____ feet

8c. The fence costs $2.50 per foot. How much will the garden fence cost?

Answer $ _____

Maneuvers with Circles © *David A. Page*

Name _____ Date _____ Class _____ 17

2. What's Inside?

 1. Figure A shows 9 squares. Each of the squares has a side length of 1 centimeter. Measure one of these squares with your ruler.

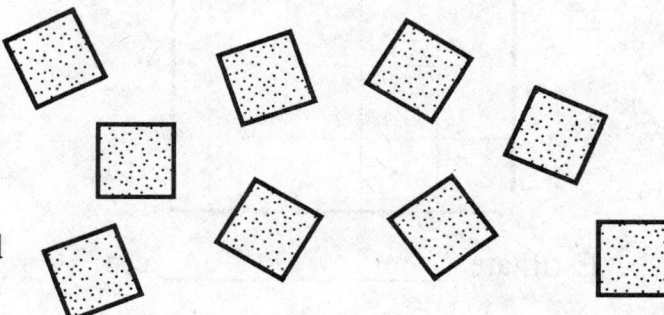

The *area* of, or space covered by, one of these centimeter squares is *1 square centimeter* and can be written as *1 cm^2*.

Figure A

These nine squares can be put together to make the larger square in Figure B. The area of the larger square is 9 cm^2.

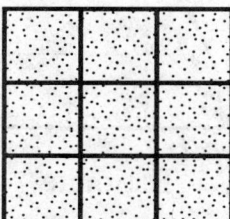

Figure B

 2. Use the 9 cm^2 square in Figure B as a comparison to estimate the area of the following figures by eye.

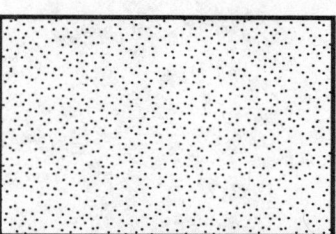 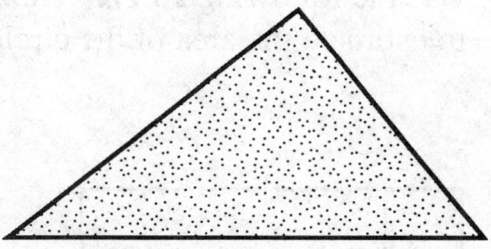

Estimate _____ cm^2 Estimate _____ cm^2

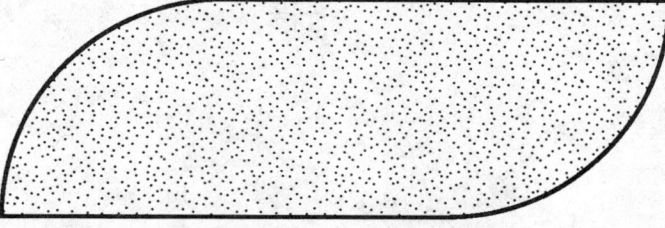 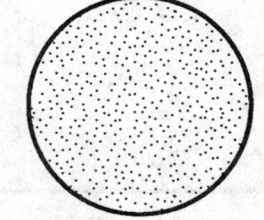

Estimate _____ cm^2 Estimate _____ cm^2

Figure C

© David A. Page   Maneuvers with Circles

18 Chapter 2

 3. Now count square centimeters and pieces of square centimeters to get a better estimate. Compare your answers with Problem 2.

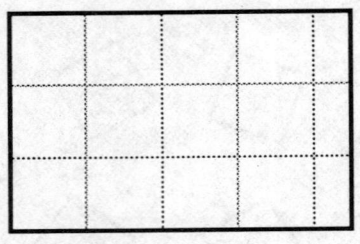   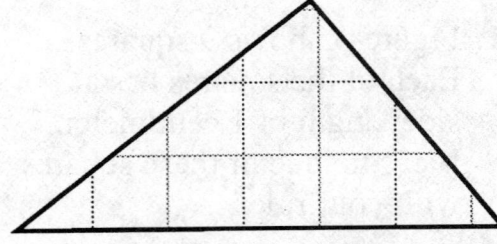

Estimate _____ cm²   Estimate _____ cm²

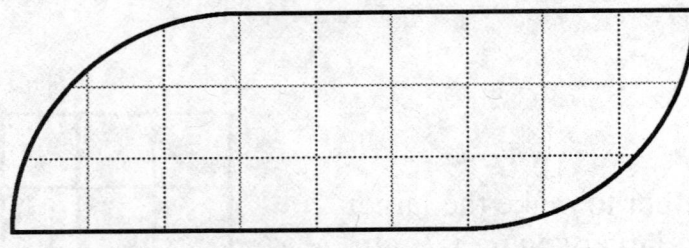   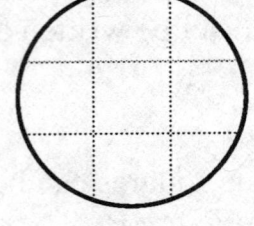

Estimate _____ cm²   Estimate _____ cm²

Figure D

 4a. Draw a circle that has a radius of 5 cm using the center mark below.

 4b. Use the following **25 cm² stamp** to estimate the area of the circle.

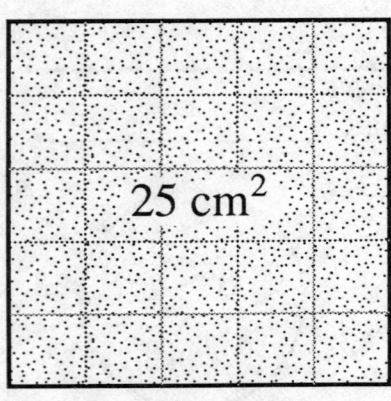

Figure E

Estimate _____ cm²

Maneuvers with Circles   © *David A. Page*

What's Inside? 19

5. A circle with a radius of 5 cm is placed in a square below. Use the following steps to find the area of the square.

5a. How many square centimeters are in one row? _____

5b. How many rows are there in the entire square? _____

5c. To find the area of the square, multiply the number of square centimeters in one row by the number of rows.

  Answer _____ cm²

| To find the area of a square, multiply its side length by itself. |

6a. Estimate the area of the circle in Figure F.

  Estimate _____ cm²

6b. Explain how you estimated the area in Problem 6a. _____
  _____

6c. Is your estimate more or less than the estimate in Problem 4b? _____
  More or Less

Figure F

© David A. Page   *Maneuvers with Circles*

The radius of the following circle is 6 cm. A shaded square is built along the radius. Notice that the side length of the square is the same length as the radius.

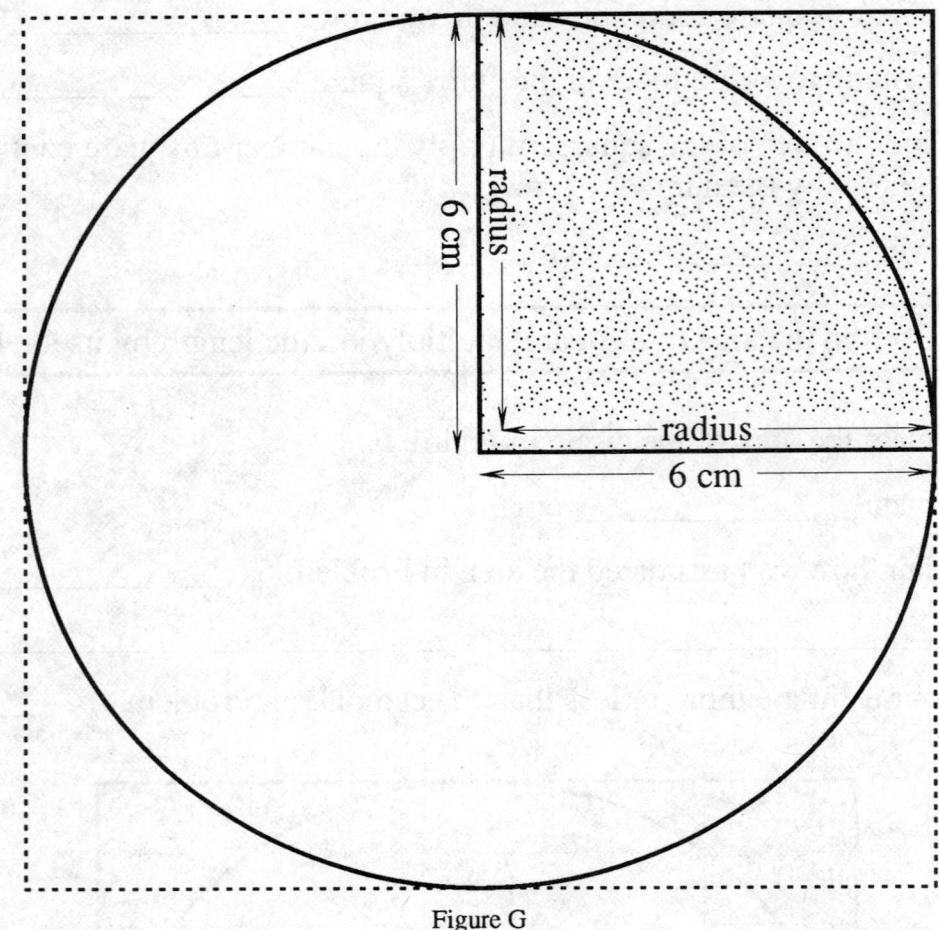

Figure G

Radius × Radius = Area of One Square Built on Radius

7. What is the area of the shaded square above? _____
Put in units.

8. Build and shade three more squares on the radii of the circle above. Figure G should look like the sketch at the right.

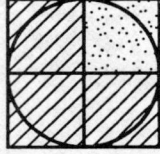
Figure H

Radius × Radius × 4 = Area of Four Squares Built on Radius

9a. What is the area of four squares? _____ cm^2

9b. Is the area of four squares more or less than the area of the circle?

Answer _____
More or Less

What's Inside?   21

Since the area of four squares is too big, let's see how three of these squares compare to the area of the circle.

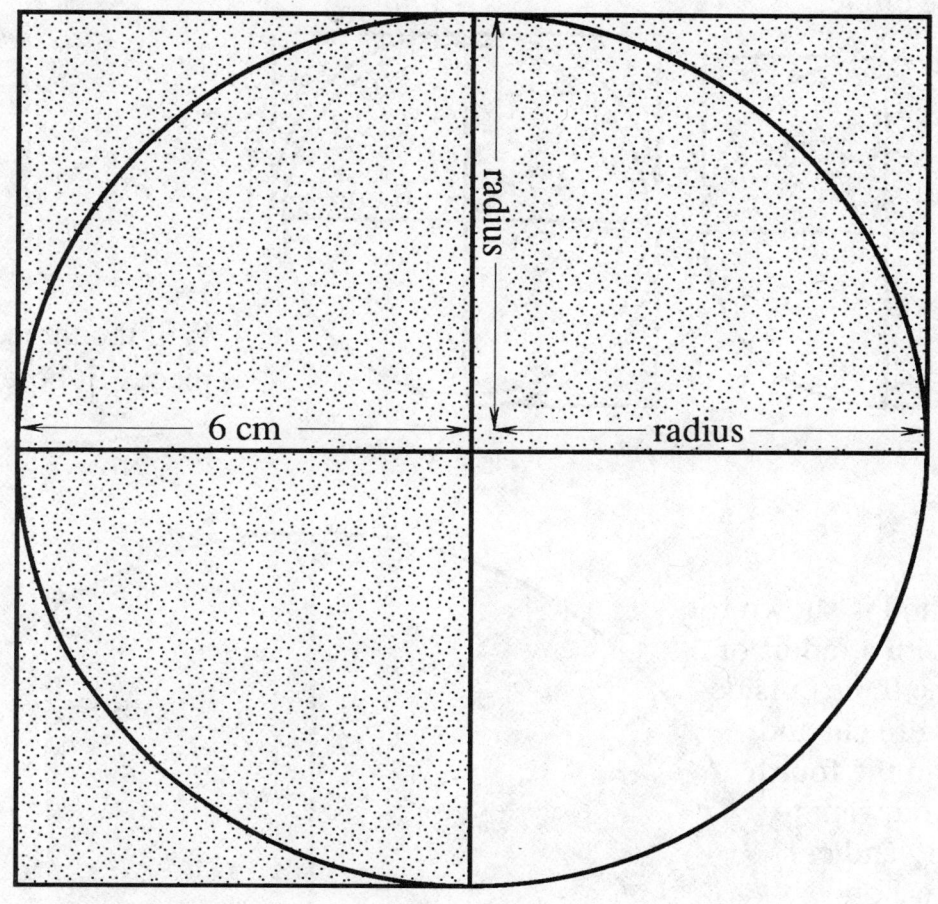

Figure J

Radius × Radius × 3 = Area of Three Squares
\underbrace{}
Area of One Square

10a. What is the area of one square? _____ cm²

10b. What is the area of three squares? _____ cm²

10c. Do you think the area of three squares is more, less, or equal to the area of the circle?

  Answer _____
         More, Less, or Equal

10d. Why? _____

_____

© David A. Page   *Maneuvers with Circles*

The following sketch shows another way to think about the area of a circle. Take the three outside pieces from the squares and see if they fit into the fourth part of the circle.

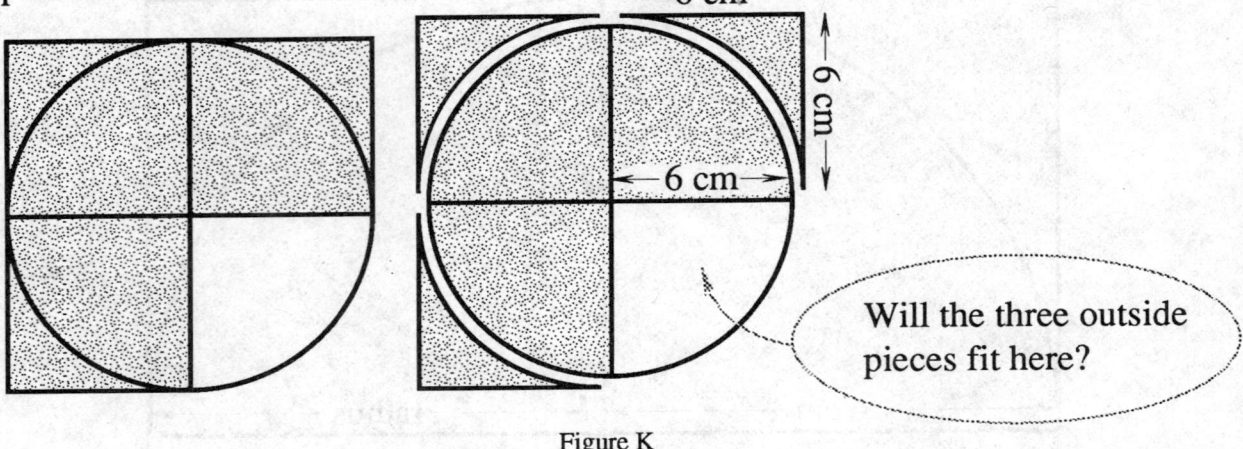

Figure K

Will the three outside pieces fit here?

This method is shown for a circle with a radius of 6 cm. The three outside corners were cut and placed into the fourth part. If you want to try it, copy and cut out Figure J on page 21.

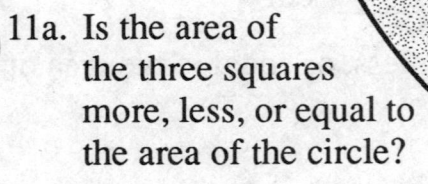

11a. Is the area of the three squares more, less, or equal to the area of the circle?

Answer _____
*More, Less, or Equal
Compare with your guess
in Problem 10c.*

Figure L

11b. Why? _____

Maneuvers with Circles © *David A. Page*

What's Inside? 23

12. You estimated the area of the circle using the following methods.

   [6] × [6] × [4] = _____ cm² is too big!

   Area of 1 square — Area of 4 squares

   [6] × [6] × [3] = _____ cm² is too small!

   Area of 1 square — Area of 3 squares

   > The area of a circle is a little more than the area of three squares built on the radii. Remember, π is a little more than 3.
   > The area of a circle is π × the area of one square built on the radius.
   > To *estimate* the area of a circle, multiply: Radius × Radius × 3.
   > To *calculate* the area of a circle, multiply: Radius × Radius × π.

13. To calculate the area of a circle with a radius of 6 cm, use the following keystrokes.

    [6] × [6] × [π] = _____ cm²

    Area of 1 square — Copy window.

14a. The circle at the right has a radius of 3 cm. Find the area of the square built on the radius.

   Area of square _____ cm²

   > Multiplying a number by itself is called *squaring* the number. Instead of multiplying 3 × 3, you can use x^2, a special key called *x-squared*.

Figure M

14b. To find the area of the square in Figure M, press [3] [x^2].

   Window _____ cm²

   Compare with Problem 14a.

14c. Estimate the area of the circle in Figure M. _____ cm²

15. Calculate the area of the circle in Figure M. List your keystrokes.

    [ ] [ ] [ ] [ ] [ ] [ ] [ ] [ ]

    You do not need to use all the keystroke boxes.

    Area _____ cm²

    Copy window.

© David A. Page — *Maneuvers with Circles*

24   Chapter 2

16a. Look at the circle at the right.
     Estimate the area of the square
     built on the radius.
     Hint: Think about the length of the radius
     rounded to the nearest whole number.

     Estimate _____ cm^2

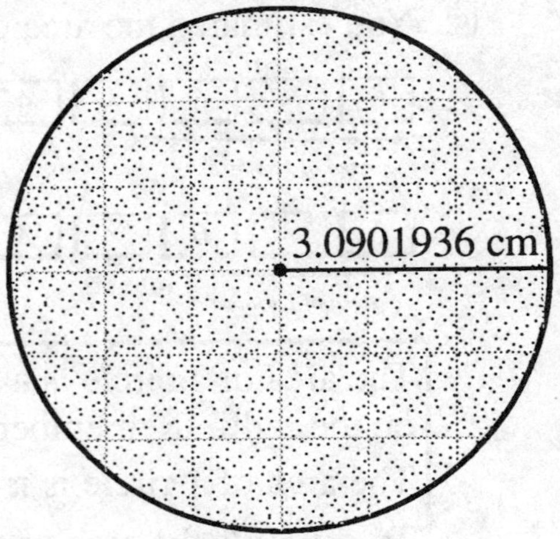

16b. Estimate the area of the circle.

     Estimate _____ cm^2

16c. Calculate the area of the circle.

     Area _____ cm^2
          You'll know.

                                                                                    Figure N

16d. List the keystrokes you used.

     | 3.0901936 | | | | | |

17. A circle has a radius of 3.9894228 cm.

17a. Estimate the area of the square built on the radius.

     Estimate _____
              Put in units.

17b. Estimate the area of the circle.

     Estimate _____
              Put in units.

17c. Calculate the area of the circle.

     Area _____ cm^2
          You'll know.

18. Calculate the area of the circle in the sketch
    at the right.

    Area  cm^2

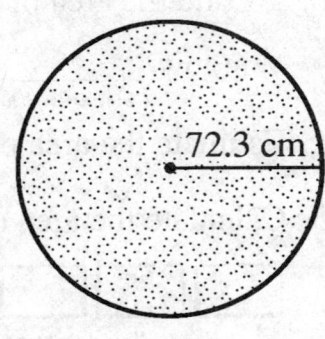

Figure P

Maneuvers with Circles   © *David A. Page*

What's Inside? 25

19a. What is the radius of the circle in Figure Q?

Radius _____ cm

19b. Calculate the area of this circle.

Area _____ cm²
R to the nearest tenth.

19c. Ramena's answer is 72.4 cm². What did she do wrong? _____

_____

Figure Q — 4.8 cm

20a. The circle in Figure R has a diameter of 5.56 cm. Draw a diameter.

20b. The following keystrokes were used by different students to calculate the area of the circle in Figure R. Only one set of keystrokes is correct. Check the correct set of keystrokes.

____ [5.56] [x^2] [×] [π] [=]

____ [5.56] [÷] [2] [x^2] [×] [π] [=]

____ [5.56] [÷] [2] [=] [x^2] [×] [π] [=]

20c. Calculate the area of the circle.

Area ☐☐ . ☐2☐ ☐9☐ ☐☐ cm²

Figure R

21. A sketch of a square picture frame is shown at the right. The frame creates a circular window for the picture. Find the area of the circular window using the following steps.

21a. What is the radius of the circle?

Radius _____ cm

21b. Area ☐☐☐ . ☐5☐ ☐5☐ ☐☐ cm²

Figure S — 27 cm

© David A. Page   *Maneuvers with Circles*

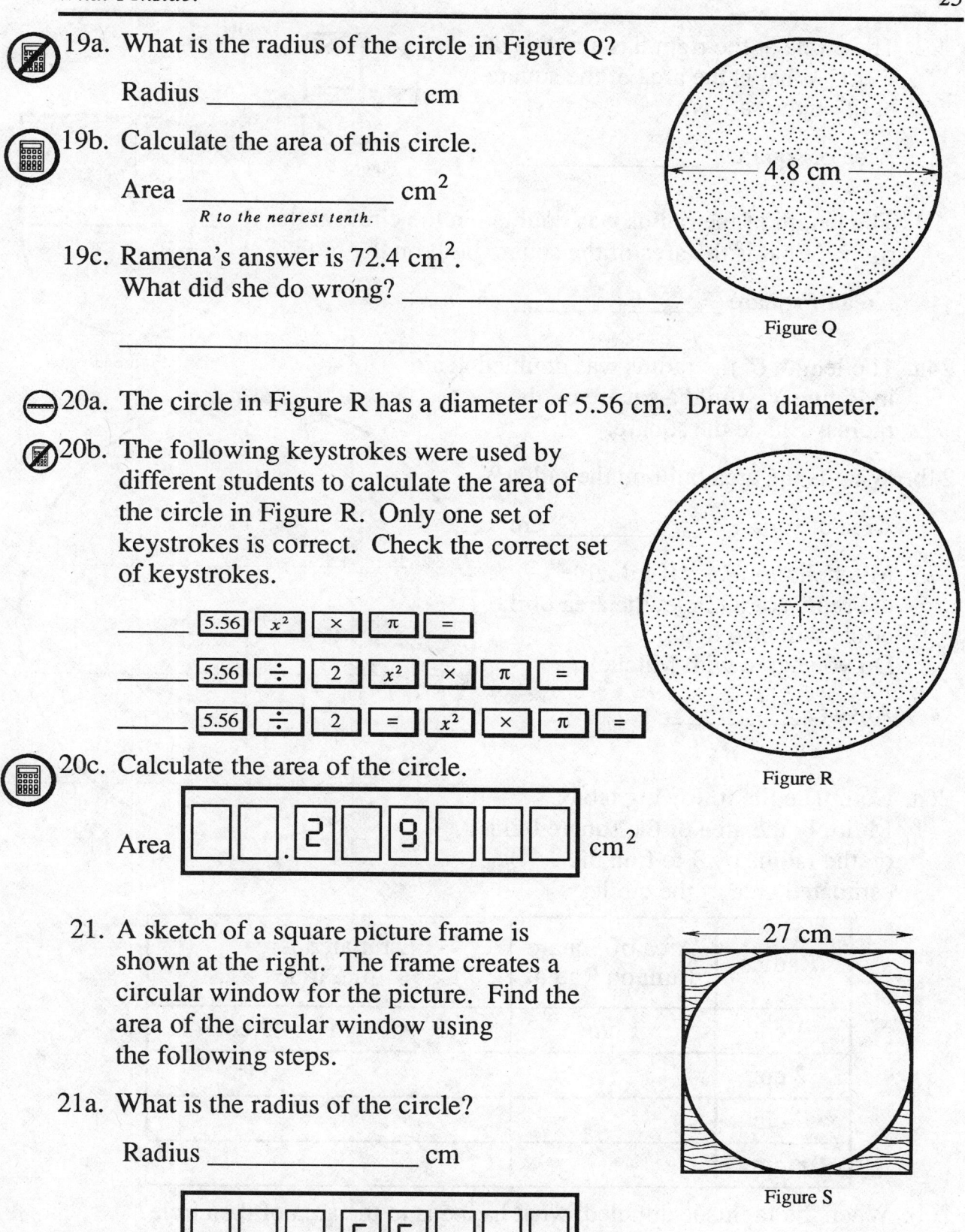

26   Chapter 2

 22. The circle at the right has a radius of
     1 cm. What is the area of the square
     built on the radius?

     Area of square _____ cm²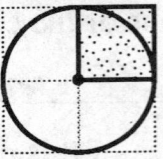
                                                   Figure T

 23. The length of the radius was doubled in the circle at the
     right. What is the area of the square built on the radius?

     Area of square _____ cm²
                                                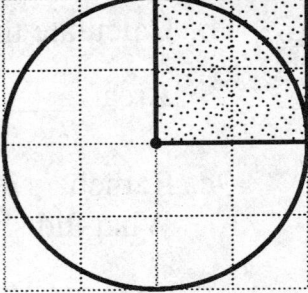
                                                   Figure U

 24a. The length of the radius was doubled again
      in Figure V. Build a square on the
      radius. Shade the square.

 24b. What is the area built on the radius?

      Area of square _____ cm²

 25. Mitchell thought, "If I double
     the radius of a circle, the area of the
     circle will also double."
     Do you agree with Mitchell?

     Answer _____
              Yes or No

 26a. Complete the following table.
      Multiply the area of the square built
      on the radius by 3 to find the
      estimated area of the circle.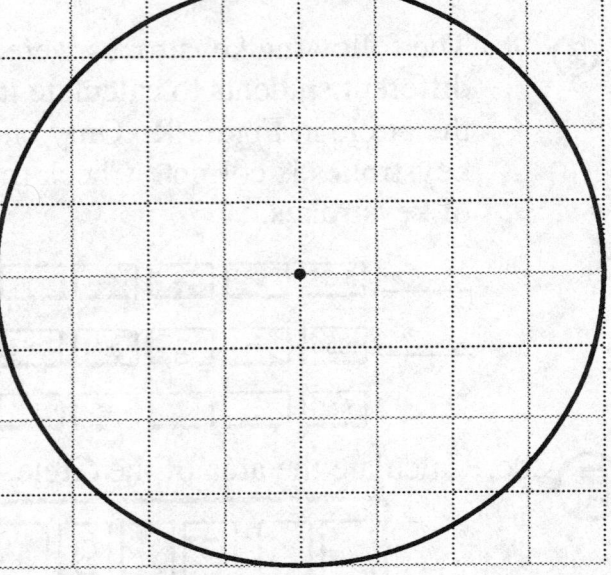
                                                      Figure V

Radius	Area of Square Built on Radius	Estimated Area of Circle
1 cm	1 cm²	3 × 1 cm² = 3 cm²
2 cm		
4 cm		
8 cm		

 26b. When the radius is doubled, what happens to the area of the circle?

 _____

Maneuvers with Circles   © *David A. Page*

Name _____ Date _____ Class _____ 27

Homework 2: What's Inside?

   1a. The figure at the right is a sketch. Find the area of the square built on the radius.

Area of square _____ cm²

1b. Estimate the area of the circle. _____ cm²

   1c. Calculate the area of the circle. _____ cm²

Compare with your estimate on page 19.

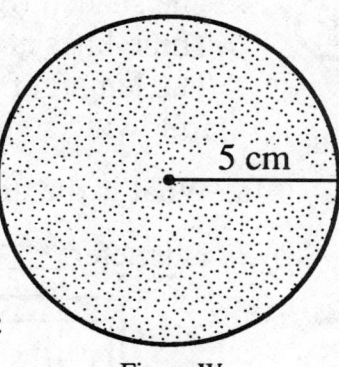

Figure W

   2. Draw a square along the radius of each circle in the following sketches. Write the area of each square on the figure. Then estimate the area of each circle.

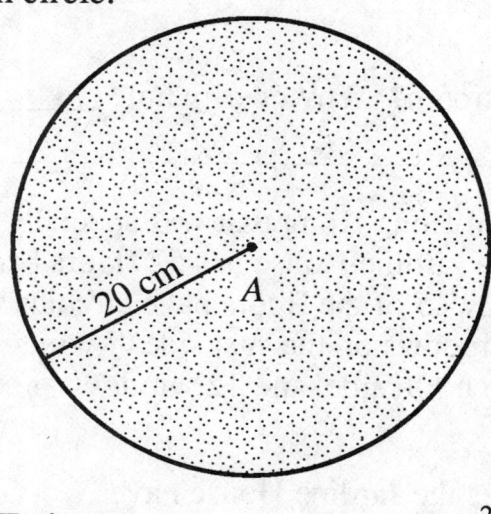 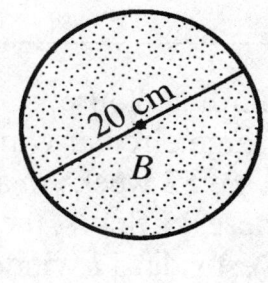

Estimate _____ cm²   Estimate _____ cm²

Figure X

   3. Calculate the area the circles in Figure X. List your keystrokes.

3a. Area of Circle A   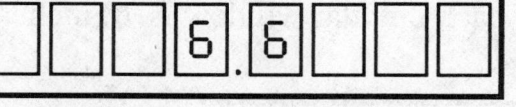 cm²

20							

You do not need to use all the keystroke boxes.

3b. Area of Circle B    cm²

20										

You do not need to use all the keystroke boxes.

© David A. Page   Maneuvers with Circles

28 Chapter 2

4. The shaded figure at the right is built from identical circles. Three lines are drawn through the centers of the circles. The length of each line is 6.45 cm. Use the following steps to find the area of the figure.

4a. What is the diameter of one circle?

Diameter _____ cm

4b. Area ☐☐.☐1☐5☐☐ cm²

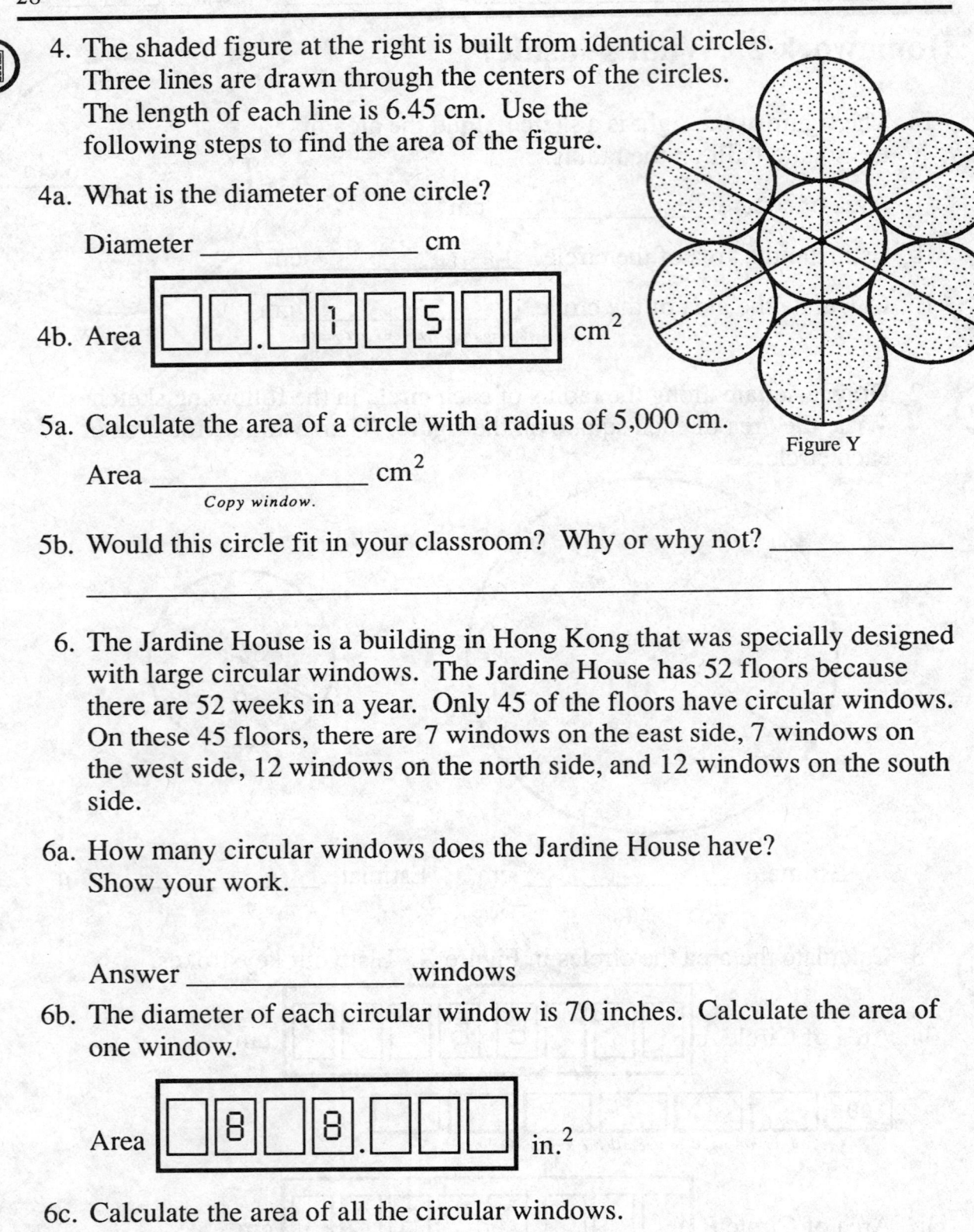

Figure Y

5a. Calculate the area of a circle with a radius of 5,000 cm.

Area _____ cm²
     Copy window.

5b. Would this circle fit in your classroom? Why or why not? _____

_____

6. The Jardine House is a building in Hong Kong that was specially designed with large circular windows. The Jardine House has 52 floors because there are 52 weeks in a year. Only 45 of the floors have circular windows. On these 45 floors, there are 7 windows on the east side, 7 windows on the west side, 12 windows on the north side, and 12 windows on the south side.

6a. How many circular windows does the Jardine House have? Show your work.

Answer _____ windows

6b. The diameter of each circular window is 70 inches. Calculate the area of one window.

Area ☐8☐8.☐☐☐ in.²

6c. Calculate the area of all the circular windows.

Area ☐☐8☐8☐☐.☐ in.²

Maneuvers with Circles © David A. Page

Name _____ Date _____ Class _____ 29

3. In and About

   1. Complete the following table. Put in units.

	Radius	Diameter	Circumference *R to the nearest hundredth.*	Area *R to the nearest hundredth.*
a.	0.25 cm			
b.		1 cm		
c.	1 cm			
d.	2 cm			
e.		8 cm		
f.	8 cm			
g.	16 cm			

2a. Draw a circle with a radius of 3 cm at the right.

2b. Look at the table and predict its circumference.

   Prediction _____ cm

2c. Look at the table and predict its area.

   Prediction _____ cm^2

2d. Calculate the circumference of this circle.

   Circumference _____ cm
   R to the nearest hundredth.

2e. Calculate the area of this circle.

   Area _____ cm^2
   R to the nearest hundredth.

2f. Compare your calculated answers with your predictions.

   3. A circle has an area of 100 cm^2. Predict its circumference.

   Prediction _____ cm

© David A. Page   *Maneuvers with Circles*

30 Chapter 3

 4. The square in the sketch at the right has a side length of 5 cm. What is the area of the square?

Area _____ cm^2

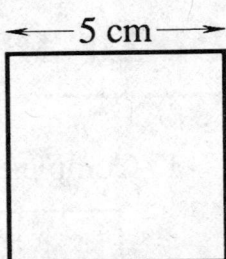
Figure A

5. The following sketch shows a square built around a circle. The area of the square is 49 cm^2.

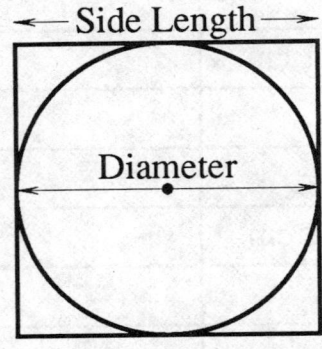
Figure B

5a. All the sides of a square are the same length. To find the side length of the square, Leo thought, "What number times itself gives me 49?" Answer Leo's question to find the side length of the square.

Side length _____ cm
_{Label the figure.}

5b. Notice that the side length of the square is the same length as the diameter of the circle. What is the diameter of the circle above?

Diameter _____ cm
_{Label the figure.}

 6a. The area of the square in the sketch at the right is 81 cm^2. Find the diameter of the circle.

Diameter _____ cm

 6b. Draw and label the length of a diameter on Figure C.

6c. What length did you find first, before the diameter?

Answer _____

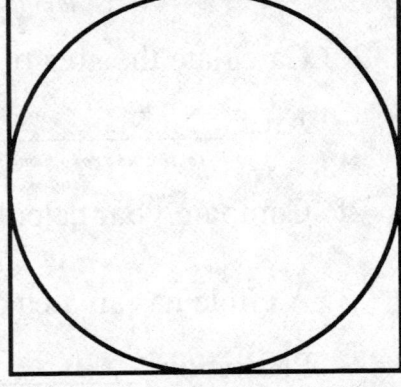
Figure C

Maneuvers with Circles © David A. Page

In and About 31

 7a. The area of the square in the sketch at the right is 70.56 cm². Circle the best estimate for the side length of the square.

6 cm   8 cm   10 cm   12 cm

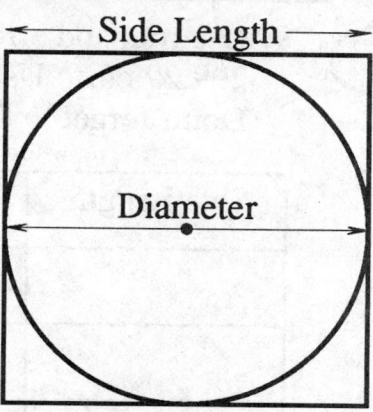

Figure D

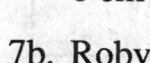 **7b.** Robyn uses trial and error to find the side length of the square. Her first estimate is too big. Complete Robyn's table. (Hint: Use x^2 to help.)

Side Length	Area *You want 70.56 cm².*
9 cm	81 cm² *Too big.*
___ cm *What's your next try?*	___ cm² *Too big or too small?*
___ cm *What's your next try?*	___ cm² *Too big or too small?*
___ cm *What's your next try?*	___ cm² *Too big or too small?*
___ cm *What's your next try?*	___ cm² *Too big or too small?*

7c. Side length of square or diameter of circle _____ cm

8a. The square in the following sketch has an area of 136.89 cm². Find the diameter of the circle (side of the square) using trial and error.

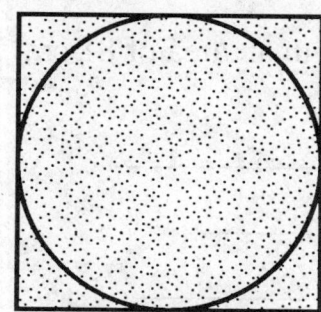

Figure E

8b. Diameter _____ cm
Label the figure.

Side Length	Area *You want 136.89 cm².*
11 cm	121 cm² *Too small.*
12 cm	144 cm² *Too big.*
___ cm *What's your next try?*	___ cm² *Too big or too small?*
___ cm *What's your next try?*	___ cm² *Too big or too small?*
___ cm *What's your next try?*	___ cm² *Too big or too small?*
___ cm *What's your next try?*	___ cm² *Too big or too small?*

© David A. Page   *Maneuvers with Circles*

32 Chapter 3

9. Use trial and error to find the side length of a square whose area is 299.29 cm². Use the following tables to keep track of your trials. Don't forget to fill in the missing units.

Side Length	Area You want 299.29 cm².

Side Length	Area You want 299.29 cm².

Side length _____ cm

There is a faster way to find the side length of a square when you know its area. Find $\sqrt{x}$ on the calculator. This is called the **square root of x**. If you know the area of a **square**, use the square root key to find its side length.

For example, the area of a square is 81 cm². To find the side length, press [81] [$\sqrt{x}$]. (On some calculators, you need to press a shift key to get $\sqrt{x}$.)

$\sqrt{81} = 9$, since $9 \times 9 = 81$. A square with an area of 81 cm² has sides that are 9 cm long.

What number times itself equals 169? _____

What is the square root of 144? _____

What is the square root of 299.29? _____
  Compare with the table in Problem 9.

$\sqrt{324}$ = _____ $\sqrt{400}$ = _____

$\sqrt{325}$ = _____ $\sqrt{410}$ = _____
  R to the nearest hundredth.   *R to the nearest hundredth.*

Maneuvers with Circles © David A. Page

In and About 33

10. The area of the square at the right is 10 cm². Estimate the side length of the square.
Think: What number squared is close to 10?

Estimate _____ cm

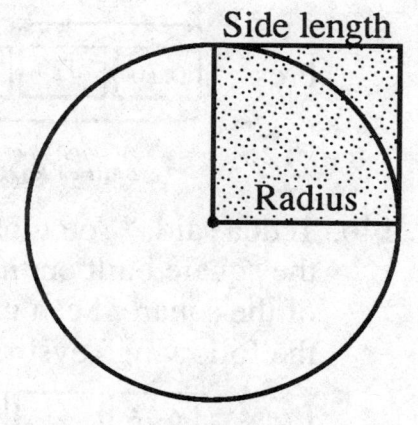

Figure F

11a. Use the square root key to calculate the side length of the square.

Press: | 10 | | √x |

Side length _____ cm
Copy window.

11b. How long is the diameter of the circle?

Diameter _____ cm
R to the nearest tenth.

12. In the sketch at the right, a square is built on the radius of a circle. The area of the square is 100 cm². Notice the side length of the square is also the radius of the circle. What is the length of the radius?

Radius _____ cm

Figure G

13. A square is built on the radius of the circle in the sketch at the right. The square has an area of 107.5369 cm².

13a. Use the answer to Problem 12 to help estimate the length of the radius.

Estimate _____ cm

13b. Find the length of the radius using √x.

Radius _____ cm

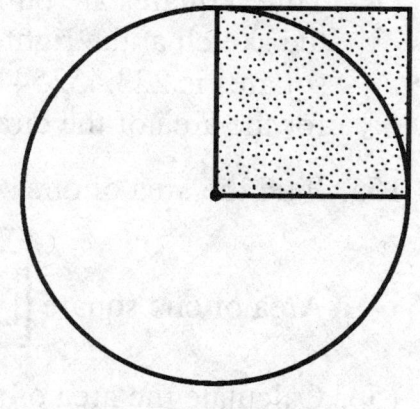

Figure H

© David A. Page   *Maneuvers with Circles*

14. A square is built on the radius in the following circle.

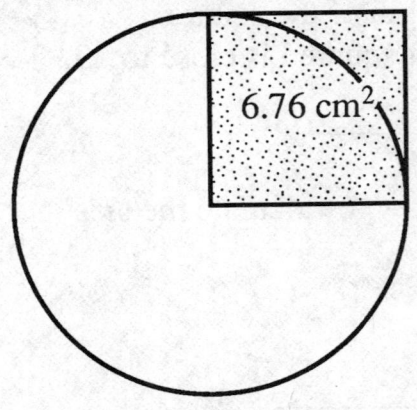

6.76 cm²

Figure J

14a. Trisha found the circle's radius and then the area of the circle using the following keystrokes.

Press: [6.76] [√x] [x²] [×] [π] [=] _____ cm²
           Radius
       ⎧‾‾‾‾‾‾‾‾‾‾⎫   Copy window.
       Area of Square
       Built on Radius

14b. Tedda said, "You took the square root and then you squared. The area of the square built on the radius is 6.76 cm². Since I already know the area of the square, I can use pi (π) to find the area of the circle." Tedda used the following keystrokes.

Press: [6.76] [×] [π] [=] _____ cm²
                         Copy window.
                         Compare with Problem 14a.

15. Three squares are built on the radius of the circle in the sketch at the right. The total area of all three squares is 223.45354 cm². This is a good estimate for the area of the circle.

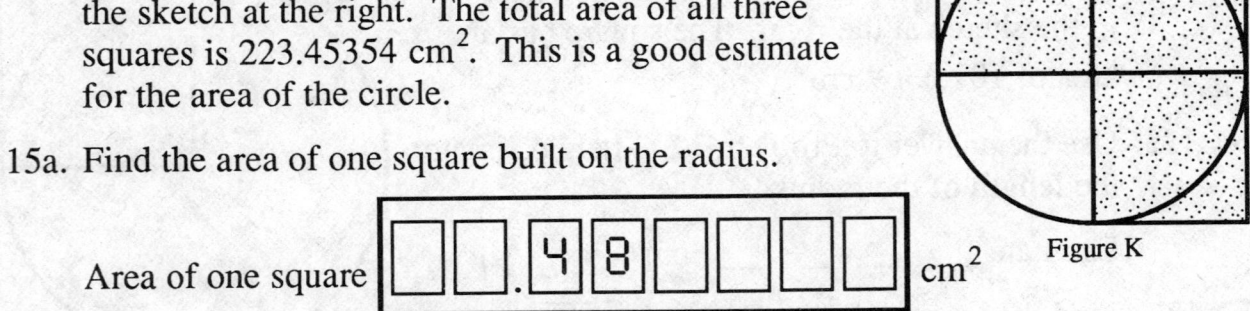

15a. Find the area of one square built on the radius.

Area of one square [ ][ ].[4][8][ ][ ][ ] cm²

Figure K

15b. Calculate the area of the circle using π. Compare your answer with the total area of the three squares.

Area _____ cm²
     You'll know.

Maneuvers with Circles   © David A. Page

In and About 35

16a. The circle in the sketch at the right has an area of 90 cm². Tammy estimated the area of the square built on the radius. She said, "The area of the circle is almost equal to the area of three squares built on the radius. I only want the area of one square."

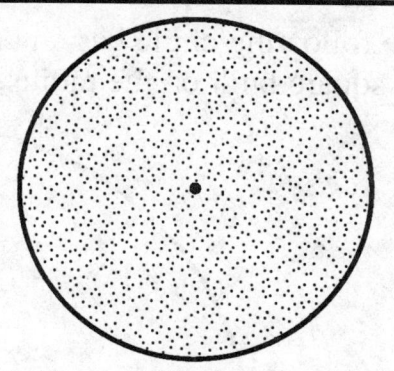

Figure L

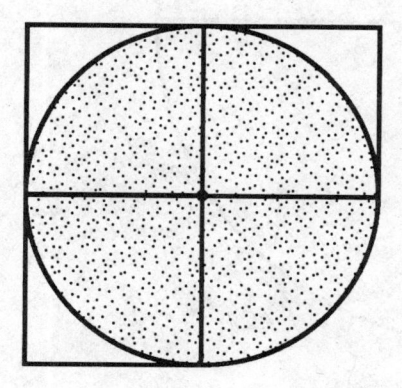

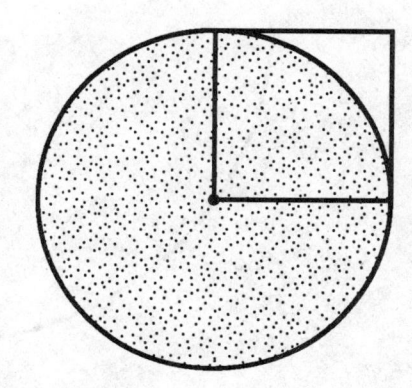

$$90 \text{ cm}^2 \div 3 = \underline{\hspace{2cm}} \text{cm}^2$$
Area of Circle   *Estimated Area of One Square*

Figure M

16b. Shannon said, "Dividing by 3 is a good estimate, but I can find a more *accurate* area of the square on the calculator." Complete Shannon's keystrokes to calculate the area of one square built on the radius.

Press: | 90 | ÷ | | = | _____ cm²
Area of Circle   *Special Number of Squares in Circle*   *Area of Square Compare with Tammy's estimate.*

16c. The side length of the square equals the length of the radius. Take the square root to find the length of the radius.

Radius _____ cm
R to the nearest tenth.

© *David A. Page*   *Maneuvers with Circles*

17a. The following circle has an area of 60.821234 cm². Estimate the area of the square built on the radius.

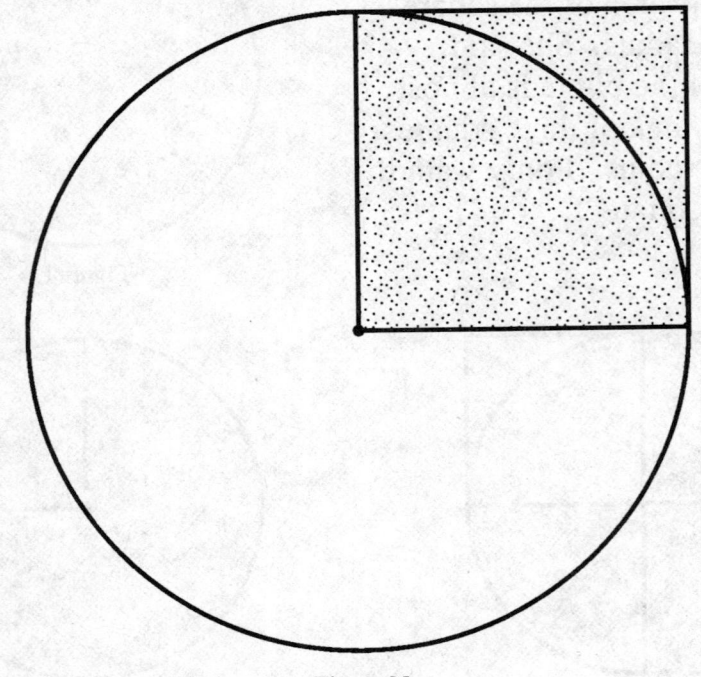

Figure N

Estimate _____ cm²

17b. Estimate the side length.

Estimate _____ cm

18a. Calculate the area of the square built on the radius in Figure N.

Area of square _____ cm²
Compare with Problem 17a.

18b. List the keystrokes for Problem 18a.

| 60.821234 | | | | |

18c. Use $\sqrt{x}$ to calculate the radius.

Radius _____ cm
Compare with Problem 17b.

18d. Measure the radius to the nearest 0.1 cm.

Radius _____ cm
Compare with Problem 18c.

In and About 37

19a. Measure the radius at the right to the nearest 0.1 cm.

   Measured radius _____ cm

19b. The area of the circle is 20 cm². What is the area of the square built on the radius?

   Area of square _____ cm²
   Copy window.

19c. What is the radius of the circle? _____ cm
   R to the nearest tenth.
   Compare with Problem 19a.

   Figure P

19d. List the keystrokes that find the length of the radius.

   | 20 | | | | | | | |

   You do not need to use all the keystroke boxes.

20. The area of the circle in the sketch at the right is 706.85835 cm². Use the following steps to calculate the circumference of the circle.

20a. What is the radius? _____ cm
   You'll know.

20b. What is the diameter? _____ cm

20c. Calculate the circumference.

   Circumference | ☐ | 4 | . | ☐ | 4 | ☐ | ☐ | ☐ | cm

   Figure Q

20d. List your keystrokes to find the circumference of the circle in one run.

   | 706.85835 | | | | | | | | | | |
   Area of Circle

21. The area of a circle is 500 cm². List the keystrokes and calculate the circumference of this circle.

   | 500 | | | | | | | | | | |

   Circumference | ☐ | ☐ | . | ☐ | 6 | 6 | ☐ | ☐ | cm

© David A. Page   *Maneuvers with Circles*

22. Measure the diameter of the circle in Figure R to the nearest 0.1 cm.

   Measured diameter _____ cm

23a. About how many diameters fit along the circumference of a circle?

   Answer _____

23b. The circumference of the circle is 24 cm. What is the exact number of diameters that fit on the circumference?

   Answer _____

Figure R

23c. Complete the following keystrokes to calculate the diameter of Figure R.

   Press: [24] [÷] [   ] [=]
           Circumference

23d. What is the length of the diameter of the circle?

   Diameter _____ cm
   Copy window.
   Compare with Problem 22.

24. A circle has a circumference of 188.49556 cm.

24a. Calculate the diameter of the circle.

   Diameter _____ cm
   You'll know.

24b. What is the radius of the circle?

   Radius _____ cm

24c. List the keystrokes you used to find the radius.

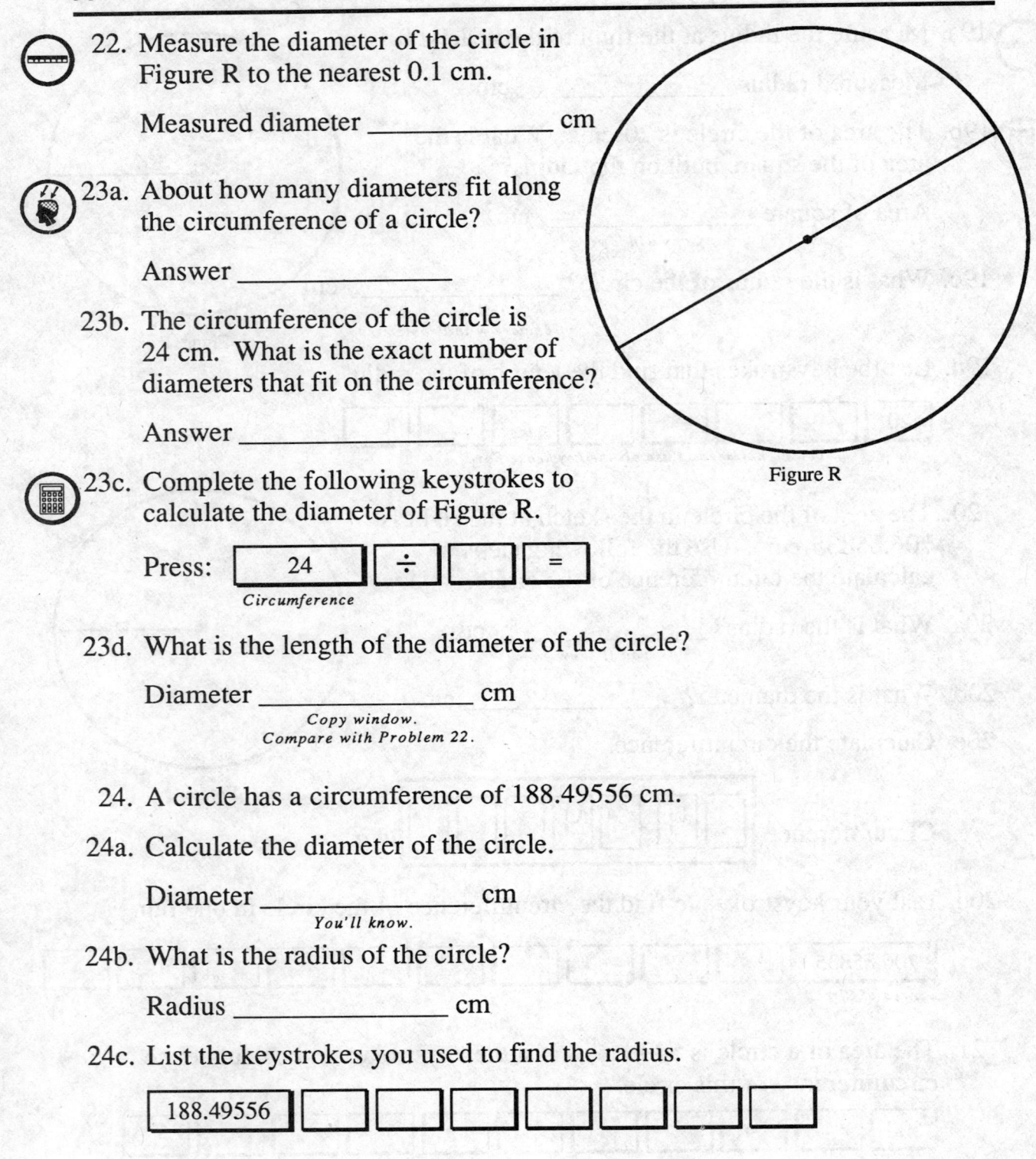

   [188.49556] [   ] [   ] [   ] [   ] [   ] [   ] [   ]

In and About 39

25. Estimate the area of the following circle.

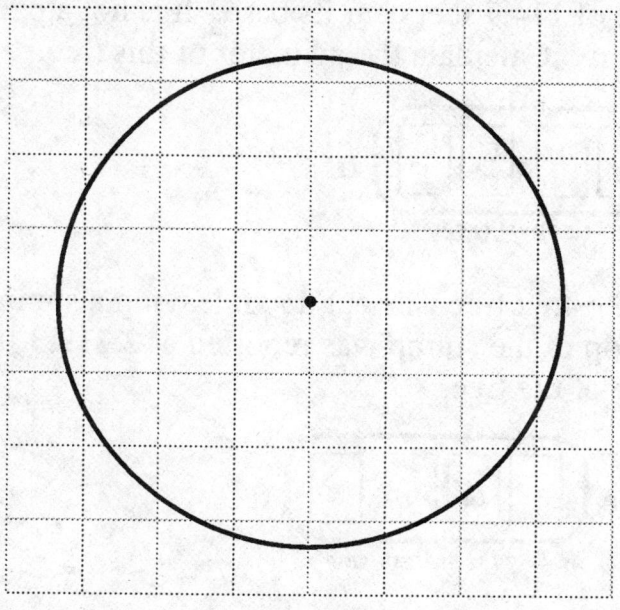

Figure S

Estimate _____ cm²

26. The circumference of the circle in Figure S is 21 cm. Use the following steps to calculate the area of the circle.

26a. What is the diameter of the circle? _____ cm
Copy window.

26b. Calculate the radius of the circle.

Radius [ ].[3][ ][2][ ][ ][ ] cm

26c. Find the area of the circle.

Area [ ][ ].[0][ ][ ][6][ ] cm²
Compare with Problem 25.

27. The circumference of a circle is 83.512488 cm. List the keystrokes to calculate the area of this circle in one run.

| 83.512488 | | | | | | | | | | |

Area _____ cm²
You'll know.

© David A. Page   *Maneuvers with Circles*

40 Chapter 3

28. The General Sherman Tree in Sequoia National Park, California has a circumference of 114.7 feet near its base. It is not a perfect circle, but assume it is here. Calculate the diameter of this tree.

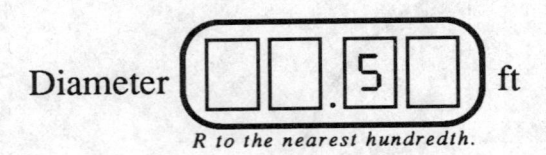

Diameter ☐☐.5☐ ft
R to the nearest hundredth.

29. A European chestnut tree was cut down leaving a stump. The area of the circle on the top of the stump was reported at 2,219.3 ft². Calculate the circumference of the tree.

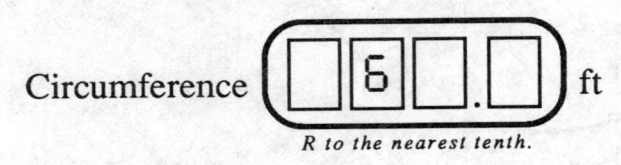

Circumference ☐6☐.☐ ft
R to the nearest tenth.

30a. Near Oaxaca, Mexico, stands an ancient cypress tree that scientists think may be 3,000 years old. The tree is about 113 feet in circumference and about 130 feet high. Calculate the diameter of this tree.

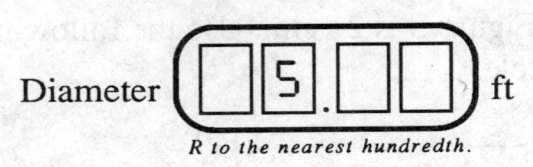

Diameter ☐5.☐☐ ft
R to the nearest hundredth.

30b. If this tree were cut, what would be the area on top of the stump?

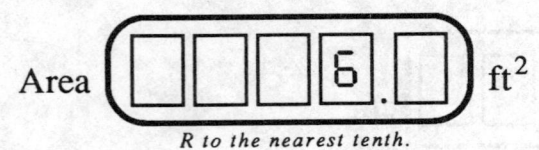

Area ☐☐☐6.☐ ft²
R to the nearest tenth.

31. Which tree in Problems 28, 29, and 30, has the largest diameter?

Answer _____

32a. Measure the circumference of a tree to the nearest whole centimeter.

Circumference _____ cm

32b. Calculate the diameter of the tree you measured. _____ cm

32c. How does this tree compare to the trees above?

_____

Maneuvers with Circles © David A. Page

Name _____ Date _____ Class _____ 41

Homework 3: In and About

1a. Yolanda's watch face has a circumference of approximately 12 cm. Estimate the diameter of her watch face.

Estimate _____ cm

1b. Calculate the diameter of the watch.

Diameter ⬚.8⬚⬚⬚8⬚ cm

2. The circumference of Elaine's pool is 48 feet. The area of Pam's pool is 255 ft². Calculate the diameter of each pool to find out which pool is larger. List the keystrokes.

2a. Diameter of Elaine's pool ⬚⬚.2⬚ ft
   R to the nearest hundredth.

2b. ⬚⬚⬚⬚⬚⬚⬚⬚
   You do not need to use all the keystroke boxes.

2c. Diameter of Pam's pool ⬚⬚.0⬚ ft
   R to the nearest hundredth.

2d. ⬚⬚⬚⬚⬚⬚⬚⬚
   You do not need to use all the keystroke boxes.

2e. Who has the larger pool? _____
   Elaine or Pam

2f. Average the diameters in Problems 2a and 2c.

Average diameter _____ ft

2g. Calculate the area of this average pool.

Area ⬚1⬚.⬚⬚ ft²
   R to the nearest hundredth.

© David A. Page *Maneuvers with Circles*

3. Marjorie's backyard is a square. The largest pool she can fit in her yard has a circumference of 40.84 ft.

3a. What is the side length of her backyard?

Length _____ ft
R to the nearest tenth.

3b. What is the area of her backyard?

Area _____ ft²
R to the nearest tenth.

3c. Draw a sketch of Marjorie's backyard. Label the sketch with the side length and area.

4a. The following rectangle is a sketch of a blackboard. Draw the largest circle within the boundaries of the rectangle.

Figure T

4b. The largest circle that can be drawn on the blackboard in Ms. Carter's class is a circle with an area of 17,671.459 cm². With this information, which can you find, the height or the width of the blackboard?

Answer _____
Height or Width

4c. Calculate this distance for the sketch above. Label the sketch.

Length _____ cm
R to the nearest whole number.

Maneuvers with Circles © *David A. Page*

In and About   43

5a. How many whole circles with a circumference of 94.24778 cm could be drawn along the height of Ms. Carter's blackboard in Problem 4?

   Answer _____ circles

5b. Sketch the circles in Figure U.

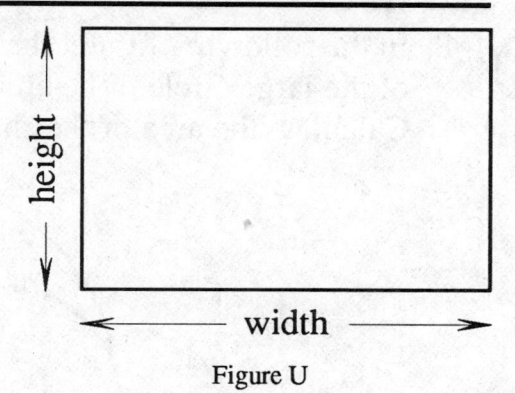

Figure U

6. The area of the square in Figure V is 38.5641 cm². Two identical circles fit inside the square. Circle an estimate for the total area of the circles.

   5 cm²

   15 cm²

   20 cm²

   25 cm²

   30 cm²

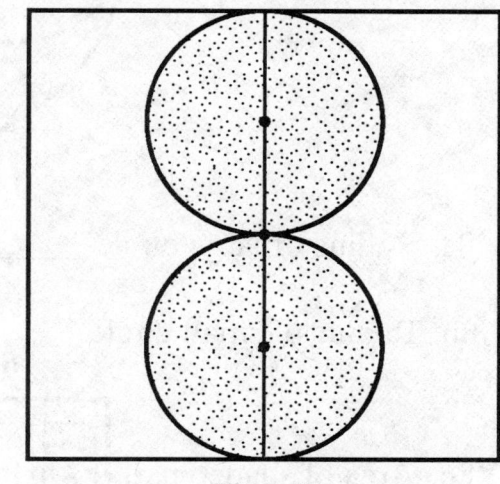

Figure V

7. Find the area of the two circles in Problem 6 using the following steps.

7a. Side length of square ☐ . 2 ☐ cm
   Measure to check.

7b. Diameter of one circle _____ cm
   Copy window.

7c. Radius of one circle ☐ . ☐ 5 ☐ ☐ cm
   Measure to check.

7d. Area of two circles ☐ ☐ . ☐ 4 4 ☐ ☐ cm²
   Compare with Problem 6.

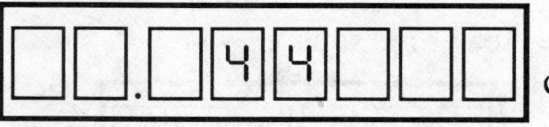

© David A. Page   *Maneuvers with Circles*

44　　　　　　　　　　　　　　　　　　　　　　　　　　　　　　　　　　　Chapter 3

8. In the following figure, the diameter of one small circle equals the radius of the large circle. The area of the large circle is 28.652582 cm². Calculate the area of the shaded part using the following steps.

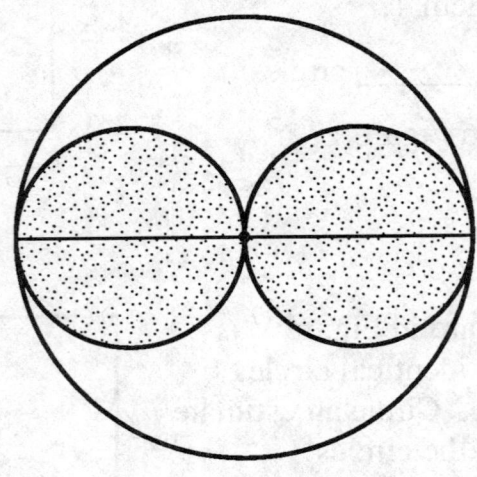

Figure W

8a. Radius of large circle _____ cm
　　　　　　　　　　　Measure to check.

8b. Radius of small circle _____ cm
　　　　　　　　　　　Measure to check.

8c. Area of shaded part ☐ ☐ . ☐ ☐ 6 2 ☐ ☐ cm²

9. The diameter of the circle at the right is the same length as the side of the square. The area of the circle is 14.186254 cm².

9a. Measure the side length of the square.

　　Measured side length _____ cm

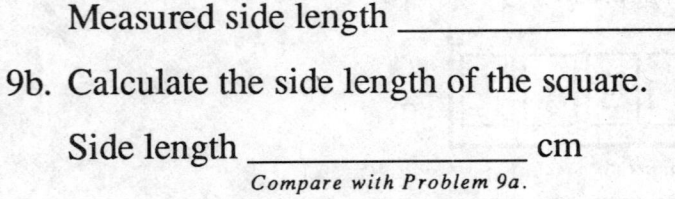

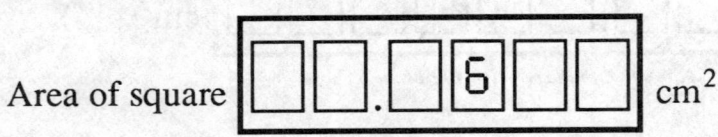

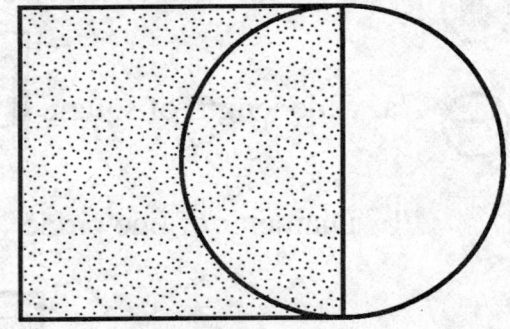

Figure X

9b. Calculate the side length of the square.

　　Side length _____ cm
　　　　Compare with Problem 9a.

9c. Calculate the area of the square.

　　Area of square ☐ ☐ . ☐ 6 ☐ ☐ cm²

Maneuvers with Circles　　　　　　　　　　　　　　　　　　　　© *David A. Page*

Name _____ Date _____ Class _____   45

4. Pieces of Circles

1. Cut out the circle below. Fold the circle in half. Cut along the fold.

    This circle is cut into two equal pieces or two half circles.
    A half circle is also called a *semicircle*.

    Since two pieces like ⌒ make a whole circle,

    the area of ⌒ equals the area of the circle

    divided by two.

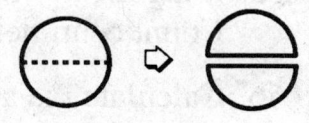

2. Glue or tape one of your semicircles at the right.

3. The area of the *entire* circle you cut out is 40 cm². What is the area of one of the semicircles?

    Glue or tape here.

    Area of semicircle _____
    Put in units.

Save the other semicircle for Problem 11.

- Cut below this line. -

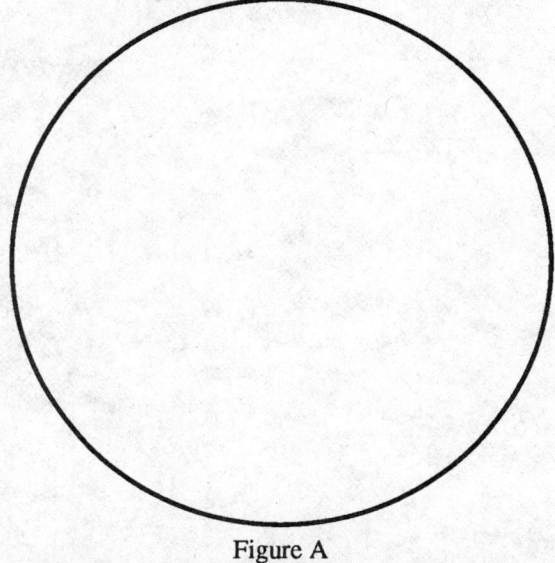

Figure A

© David A. Page   *Maneuvers with Circles*

46   Chapter 4

4. Count shaded square centimeters to estimate the area of the semicircle at the right.

   Estimate _____
   Put in units.

5. Use the following steps to calculate the area of the semicircle.

   Figure B

5a. Imagine the entire circle as shown in Figure C.
    Count centimeters to find the radius of the circle. _____ cm

5b. Calculate the area of the *whole* circle using the following keystrokes. Keep the answer in the window.

   Press: [3] [x^2] [×] [π] [=]

   Answer _____ cm^2
   Copy window.

5c. Calculate the area of the semicircle.

   Area of semicircle _____ cm^2
   R to the nearest whole number.
   Compare with your estimate in Problem 4.

   ←— radius —→

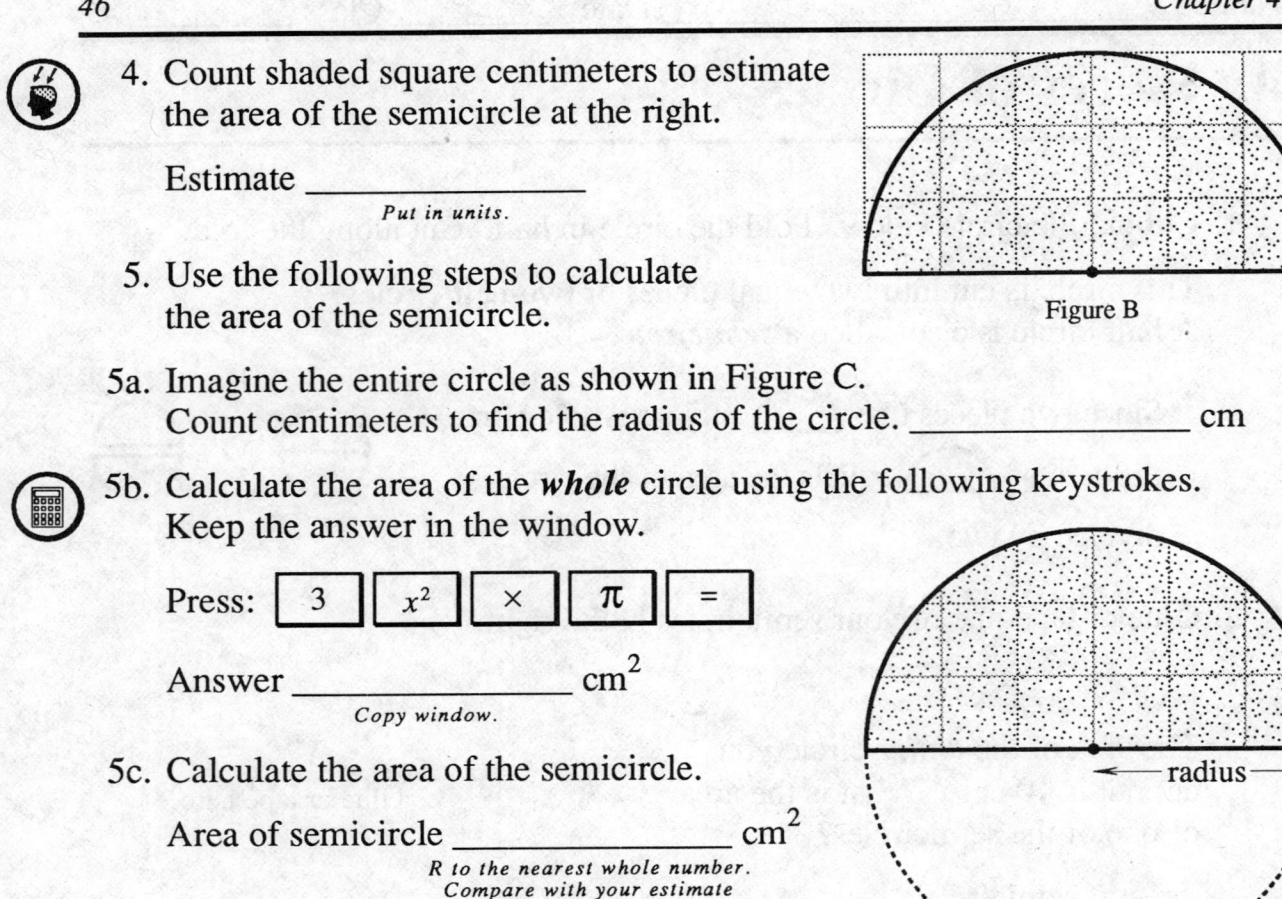

   Figure C

5d. List the keystrokes that calculate the area of a semicircle with a radius of 3 cm.

   [3] [ ] [ ] [ ] [ ] [ ] [ ] [ ] [ ] [ ] [ ]
   You do not need to use all the keystroke boxes.

Maneuvers with Circles   © *David A. Page*

Pieces of Circles

6a. Measure the radius of the semicircle at the right to the nearest whole centimeter.

Radius _____
Put in units.

6b. Calculate the area of a circle with this radius. Keep the answer in the window.

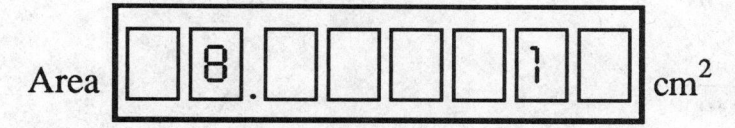

Area | |8|.| | | |1| | cm²

6c. Calculate the area of the semicircle.

Area | | |.|2|6| | | | cm²

6d. Round your answer to the nearest tenth.

Area _____
Put in units.

Figure D

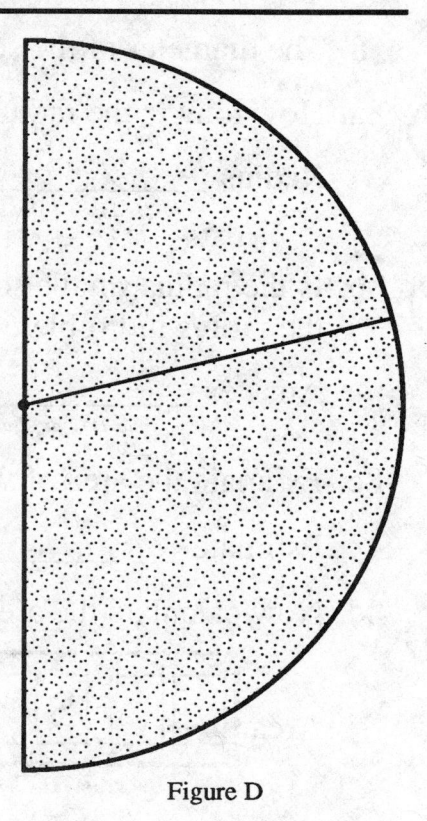

7. The radius of the following semicircle is 4 cm. Calculate the area of the semicircle. List your keystrokes.

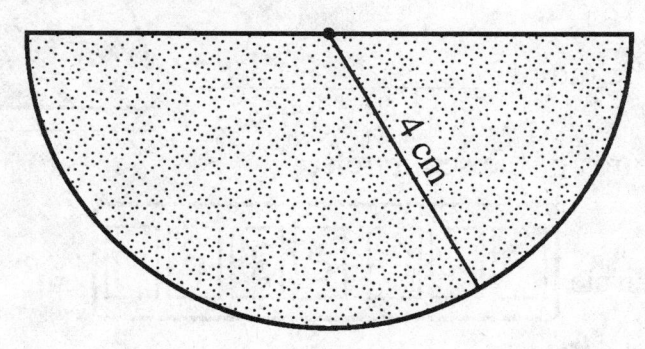

4 cm

Figure E

Area of semicircle | | |.|3|2| | | | cm²

| | | | | | | | | |

8. The diameter of the semicircle at the right is 7.4 cm.

8a. How long is the radius?

Radius _____ cm
Measure to check your answer.

8b. Is the radius smaller or larger than the radius in Problem 7?

Answer _____
Smaller or Larger

8c. Estimate the area of the semicircle.

Estimate _____ cm²

8d. Now calculate the area of the semicircle.

Area [ ][ ].[5][4][ ][ ] cm²
Compare with your estimate.

Figure F

9. The semicircle in the sketch at the right has a diameter of 18 cm.

9a. How long is the radius? _____
Put in units.

9b. Calculate the area of the semicircle.

Area of semicircle [ ][ ][ ].[2][3][ ] cm²

Figure G

9c. List your keystrokes.

18					

You do not need to use all the keystroke boxes.

10. The diameter of a semicircle is 17.697936 cm. Calculate the area of the semicircle.

Area of semicircle _____ cm²
The sum of the digits is 6.

Maneuvers with Circles © *David A. Page*

Pieces of Circles

11a. Take your other semicircle from page 45. Fold the semicircle in half. Cut along the fold.

Since four pieces like ◠ make a whole circle, the area of ◠ equals the area of the circle divided by four.

◠ is called a ***quarter-circle***.

Tape or glue quarter-circle here.

11b. The area of the entire circle from page 45 is 40 cm^2. What is the area of one quarter-circle?

Area of quarter-circle _____

Put in units.

12. Estimate the area of the quarter-circle at the right.

Estimate _____ cm^2

13. Use the following steps to calculate the area of the shaded quarter-circle.

Figure H

13a. Calculate the area of the whole circle in Figure H. Keep the answer in the window.

Area of circle _____ cm^2

Copy window.

13b. Four quarter-circles build the circle. The quarter-circle is one piece out of four. What should you do to find the area of the quarter-circle?

Answer _____

13c. Calculate the area of the quarter-circle.

Area of quarter-circle _____ cm^2

Copy window.

Compare with your estimate in Problem 12.

© David A. Page   *Maneuvers with Circles*

50   Chapter 4

14a. The circle in the sketch at the right has a radius of 3.8 cm. Calculate the area of the **whole** circle.

Area ☐☐.☐☐45☐☐ cm²

14b. Calculate the area of the shaded quarter-circle.

Area of quarter-circle ☐☐.3☐ cm²

R to the nearest hundredth.

⟵ 3.8 cm ⟶

Figure J

15a. Draw a circle with a radius of 3.5 cm at the right.

15b. Draw two diameters to divide the circle into quarter-circles. Shade one of the quarter-circles.

15c. Is the area of this quarter-circle smaller or larger than the one in Problem 14?

Answer _____
Smaller or Larger

15d. Calculate the area of the shaded quarter-circle.

Area of quarter-circle ☐.6☐ cm²

R to the nearest hundredth.

16. The radius of the quarter-circle at the right is 3.3 cm.

⟵ 3.3 cm ⟶

16a. Compare the quarter-circles in Problems 14, 15, and 16. Estimate the area of the quarter-circle in Figure K.

Estimate _____
Put in units.

16b. Calculate the area of the quarter-circle.

Area ☐.5☐2☐☐ cm²

Figure K

16c. List your keystrokes.

| 3.3 | ☐ | ☐ | ☐ | ☐ | ☐ | ☐ | ☐ | ☐ |

Maneuvers with Circles   © *David A. Page*

Pieces of Circles 51

Pieces Put Together

1. Three quarter-circles are put together to build the shaded figure at the right. Count to estimate the area of the shaded figure.

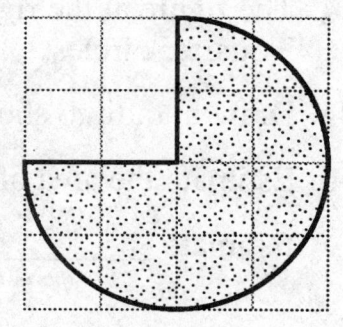

    Estimate _____
    Put in units.

    Figure L

2. Calculate the area of the shaded figure using the following steps.

2a. Calculate the area of the circle. Do not clear the window.

   Area of circle ☐☐.☐66☐☐☐ cm²

2b. Calculate the area of one quarter-circle. Do not clear the window.

   Area of quarter-circle ☐.1☐1☐☐☐ cm²

2c. Since there are three quarter-circles, multiply by 3 to find the area of the shaded figure.

   Area of shaded figure _____ cm²
   Copy window.
   Compare with Problem 1.

2d. List the keystrokes to find the area of the shaded figure in one run.

   ☐ ☐ ☐ ☐ ☐ ☐ ☐ ☐ ☐ ☐

3. Kendra used a different method and pressed the following keystrokes.

   | 2 | x^2 | × | π | − | 2 | x^2 | × | π | ÷ | 4 | = |

3a. Put a loop around the keystrokes that find the area of the whole circle.

3b. Put a box around the keystrokes that find the area of one quarter-circle.

3c. Explain Kendra's method. _____

© David A. Page   *Maneuvers with Circles*

52   Chapter 4

4. The figure at the right is built from
   3 quarter-circles.

4a. Draw in radii to show the 3 quarter-circles.

4b. Estimate the area of the ⸻ shading.

   Estimate _____
             Put in units.

5a. Calculate the area of the ⸻ shading.

   Area _____ cm²
        R to the nearest whole number.
        Compare with your estimate.

                                                                                    Figure M

5b. List your keystrokes for Problem 5a.

   | 3 | | | | | | |
   | | | | | | | |
   You do not need to use all the keystroke boxes.

6. The figure at the right is built from quarter-circles.
   Calculate its area. Show your work.

   ← 2.5 cm

   Area ☐ ☐ . ☐ 2 6 ☐ ☐ ☐ cm²

                                                                                    Figure N

7. The sketch at the right is built from
   quarter-circles. The diameter is 16 cm.
   Calculate the area of the figure.

   Area ☐ 5 ☐ . ☐ ☐ cm²
        R to the nearest hundredth.

                                                                                    Figure P

Maneuvers with Circles   © *David A. Page*

Name _____ Date _____ Class _____ 53

Homework 4: Pieces of Circles

1. The semicircle in the following sketch has a radius of 28.028357 cm. Calculate the area of the semicircle.

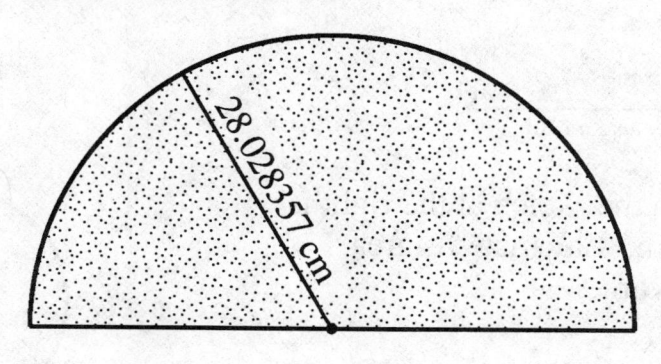

Figure Q

Area _____ cm²
The sum of the digits is 10.

2. The semicircle in the following sketch has a diameter of 55 cm.

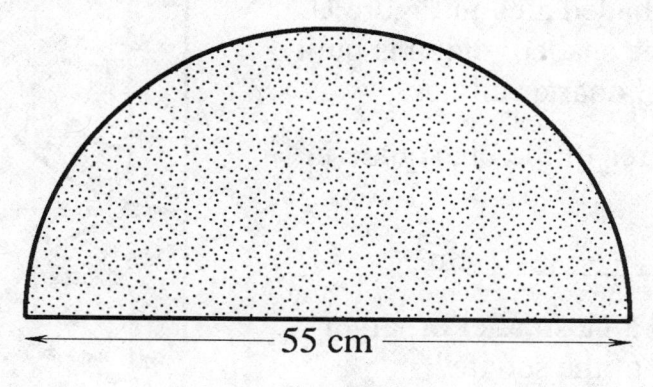

←——— 55 cm ———→
Figure R

2a. Is the area of this semicircle smaller or larger than the area in Problem 1?

Answer _____
Smaller or Larger

2b. Explain your answer. _____

_____

2c. Calculate the area of the semicircle.

Area _____ cm²
R to the nearest tenth.

© David A. Page   Maneuvers with Circles

54 Chapter 4

3. Calculate the area of the quarter-circle at the right. List your keystrokes. Draw your own keystroke boxes.

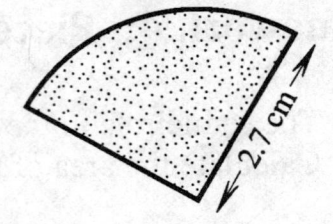

Figure S — 2.7 cm

Area _____ cm²
R to the nearest tenth.

4. The radius in the sketch at the right is 6.5147002 cm. Calculate its area. Show your work.

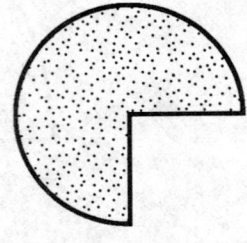

Figure T

Area _____ cm²
You'll know.

5. Look at the shaded area in Figure U. Notice that the shaded region is built from identical quarter-circles.

5a. What is the area of the big square in Figure U?

Area _____ cm²

5b. Is the shaded area smaller or larger than the area of the square?

Answer _____
Smaller or Larger

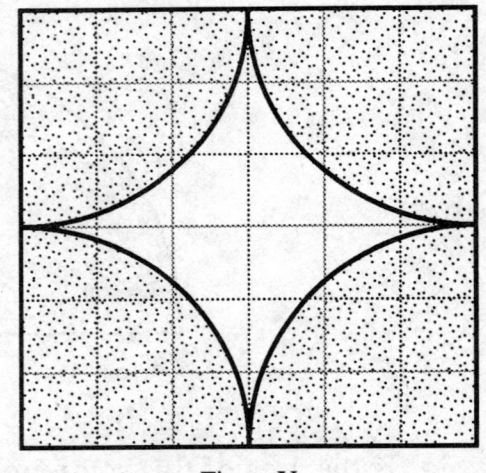

Figure U

5c. Why? _____

_____

6. Calculate the area of the shaded region in Figure U. Hint: How many quarter-circles are there?

Area ☐☐.☐☐4 3☐☐ cm²

Maneuvers with Circles © David A. Page

Name _____ Date _____ Class _____ 55

5. Remember That Number

1. The circle at the right has a diameter of 6.3661978 cm.

    1a. Calculate the circumference of the circle. List your keystrokes.

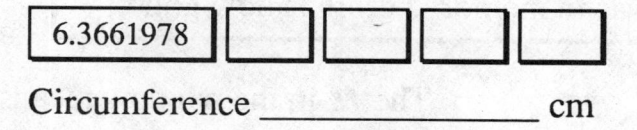

    | 6.3661978 | | | | |

    Circumference _____ cm
    You'll know.

    6.3661978 cm

    1b. Calculate the area of the circle. List your keystrokes.

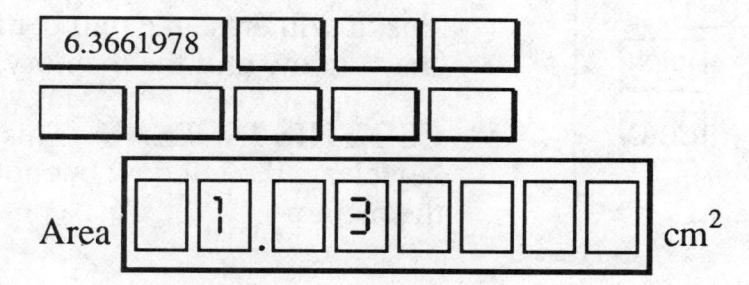

    | 6.3661978 | | | |
    | | | | |

    Area | 1 | . | 3 | | | | cm²

    Figure A

In the problems above, 6.3661978 was pressed twice. Instead of pressing a messy number twice, use the calculator's scratch pad (called **memory**) to save a number. Memory is very helpful with complicated problems.

On the TI-30 SLR+™, the four keys at the bottom left run the memory.

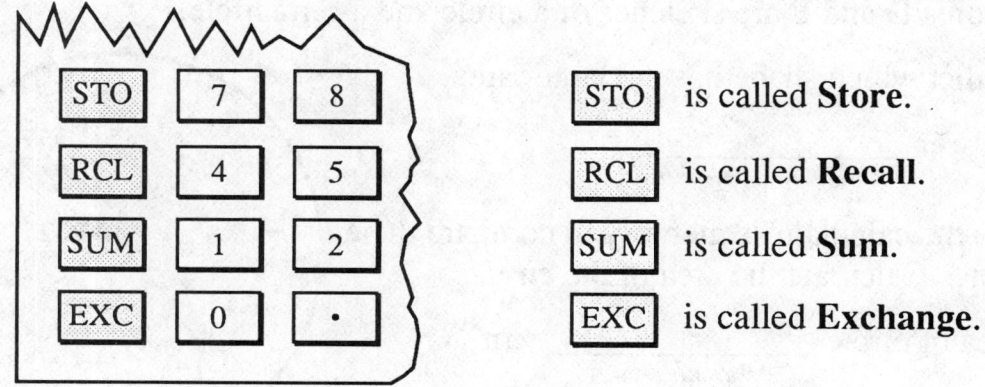

| STO | 7 | 8 | STO is called **Store**.
| RCL | 4 | 5 | RCL is called **Recall**.
| SUM | 1 | 2 | SUM is called **Sum**.
| EXC | 0 | . | EXC is called **Exchange**.

Figure B

The following pages describe how to use these four keys.

© *David A. Page*   *Maneuvers with Circles*

56   Chapter 5

$\boxed{\text{STO}}$ Key

One way to activate your calculator's memory is to use the $\boxed{\text{STO}}$ key. Put the year you were born into your calculator. For example, if you were born in 1980, press $\boxed{1980}$. Now press $\boxed{\text{STO}}$. The year you were born is now in the calculator's memory.

> $\boxed{\text{STO}}$ takes a copy of the number in the window and puts it into the memory. Any number that was in memory before is now gone.

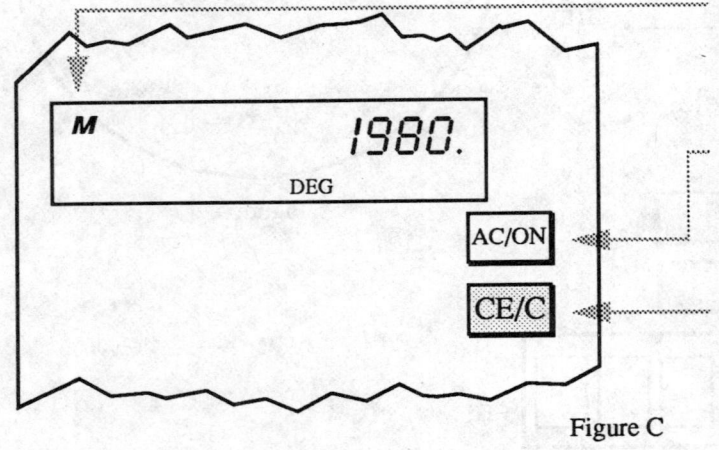

The **M** in the window means you have a number in **M**emory.

Display shows: **M** 1980. DEG

WARNING! If you press this, it will erase the number in the window **and** the memory! (AC/ON)

USE THIS INSTEAD. This key clears the window but not the memory. (CE/C)

Figure C

$\boxed{\text{EXC}}$ Key

The $\boxed{\text{EXC}}$ key exchanges the number in the window with the number in the memory.

2. Figures D and E are sketches of a circle and a semicircle.

2a. Predict which shape has the larger area.

Answer _____
    Circle or Semicircle

2b. Use the calculator's memory to compare their areas. Calculate the area of the *circle*.

Area of circle _____ cm²
            You'll know.

Press $\boxed{\text{STO}}$. Do not press $\boxed{\text{AC/ON}}$.

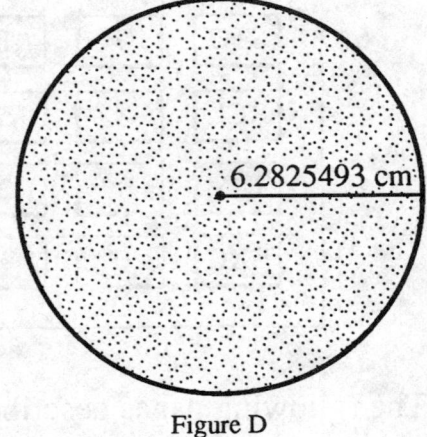

Figure D (circle with radius 6.2825493 cm)

Maneuvers with Circles   © David A. Page

Remember That Number 57

2c. Calculate the area of the *semicircle*.

Area of semicircle _____ cm²
You'll know. Not 250.

Press EXC to see the circle's area.

You EXChanged two numbers. The number in the memory went into the window. The number that was in the window went into the memory.

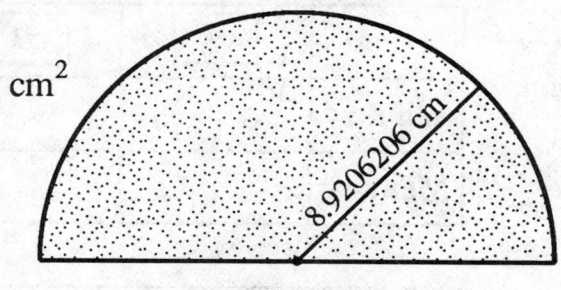

Figure E — semicircle with radius 8.9206206 cm

2d. Which area is larger? _____
Circle or Semicircle

2e. Press EXC again. Now the semicircle's area is in the window and the circle's area is in memory. Press AC/ON. You just cleared the window and the memory. The **M** is gone. "0" is in the window and memory. Press EXC to check.

Some people call EXC the "swap key." One important use of EXC is to take a peek at the memory without losing the number in the window. Press EXC once to peek at what is in memory. Press EXC again to get the numbers back the way they were.

Complete the tables in Problems 3 through 6. After you press each key, fill out what is in the window and memory.

3.
Press:	AC/ON	27	STO	5
Window	0	27		
Memory	0			

4.
Press:	AC/ON	88	STO	121	EXC
Window					
Memory					

© David A. Page   *Maneuvers with Circles*

5.

Press:	AC/ON	53	STO	47	STO
Window					
Memory					

6.

Press:	AC/ON	8.741	STO	7	EXC	EXC	EXC
Window							
Memory							

Press the keystrokes in the following problems. Then answer the questions.

7. Press: AC/ON 125 STO + 25 =

7a. What do you see in the window? _____

7b. What number is in memory? _____

Peek to check.

8. Press: AC/ON 73 × EXC =

8a. What do you see in the window? _____

8b. What number is in memory? _____

9. Fill in the following boxes to create your own problem using STO and EXC.

Press: AC/ON ☐ ☐ ☐ ☐ ☐ ☐ ☐

9a. What number do you see in the window? _____

9b. What number is in memory? _____

Remember That Number

RCL Key

10a. The circle at the right has a diameter of 7.0028175 cm. Store the diameter in memory.

   Press: [7.0028175] [STO]

10b. The diameter is now in the window and in the memory. Calculate the circumference of the circle.

   Press: [×] [π] [=]

   Circumference _____ cm
   _{*You'll know.*}

7.0028175 cm

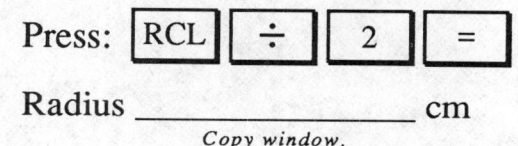

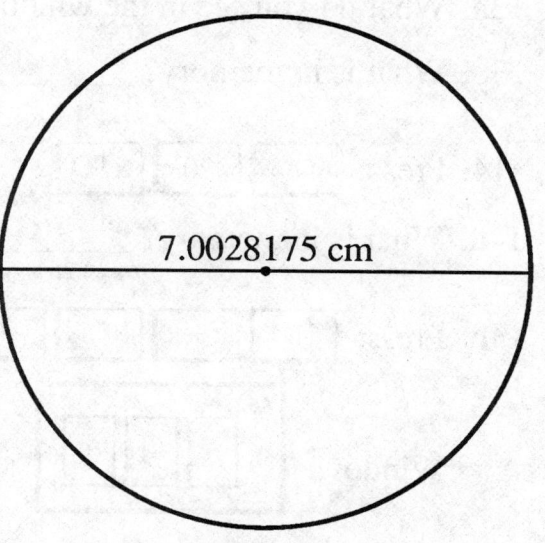

Figure F

10c. Find the radius of the circle.

   Press: [RCL] [÷] [2] [=]

   Radius _____ cm
   _{*Copy window.*}

10d. When you pressed [RCL], what number came to the window? _____

> [RCL] (Recall) takes a copy of the number in memory and puts it in the window. The number that used to be in the window is gone.

Complete the following tables.

11.

Press:	AC/ON	100	STO	23	RCL
Window					
Memory					

12.

Press:	AC/ON	103	STO	406	RCL
Window					
Memory					

© David A. Page

Maneuvers with Circles

60   Chapter 5

13. Press: [AC/ON] [20] [STO] [65] [+] [35] [=] [×] [RCL] [=]

13a. What do you see in the window? _____

13b. What is in memory? _____

14. Press: [AC/ON] [20] [STO]

14a. What is in memory? _____
Don't clear your calculator.

14b. Press: [180] [+] [RCL] [+] [RCL] [+] [RCL] [=]

Window: | M 4 . |

14c. What number is being added to the window every time you press [+] [RCL] in Problem 14b?

Answer _____

[RCL] vs. [EXC]

15. Fill in the following tables. Press [EXC] twice to peek into memory.

15a.
Press:	25	STO	30	EXC
Window				
Memory				

15b.
Press:	25	STO	30	RCL
Window				
Memory				

15c. The keystrokes in the two tables above only differ by the last keystroke. What is the difference between [EXC] and [RCL]?

_____

Maneuvers with Circles   © *David A. Page*

Remember That Number 61

16. Press: | 2 | x^2 | × | π | = | STO |

| 2.8 | x^2 | × | π | ÷ | 2 | = |

16a. The top row finds the area for which shape? _____
                                                                                                 Circle or Semicircle

16b. The bottom row finds the area for which shape? _____
                                                                                                  Circle or Semicircle

16c. Is the answer to the top row in the window or memory? _____
                                                                                                  Window or Memory

16d. If you want to compare the two areas, you don't want to lose either one. Which key do you press, EXC or RCL? _____

16e. Which area is larger? _____

SUM Key

> Each time you press SUM, it takes the number that is in the window and adds it to the number in memory.

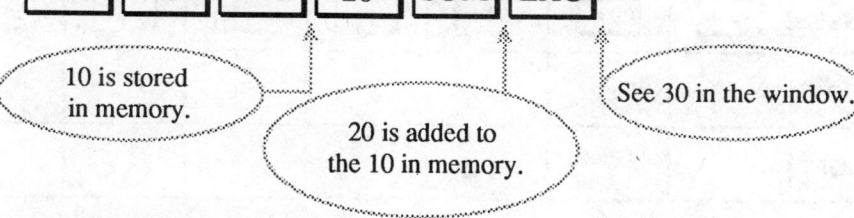

17. Press: | AC/ON | 10 | STO | 20 | SUM | EXC |

- 10 is stored in memory.
- 20 is added to the 10 in memory.
- See 30 in the window.

17a. What is in the window? _____

17b. What is in memory? _____

18. Press the keystrokes listed in the following table.

- 395 is stored in memory.
- 5 is added to memory each time SUM is pressed.

Press:	AC/ON	395	STO	5	SUM	SUM	SUM	EXC
Window	0	395						
Memory	0	0			400			

© David A. Page   *Maneuvers with Circles*

Each table below has a series of keystrokes.

a. First predict what will be in the window and memory at the end of each series. Write your predictions next to each table.

b. Then use your calculator to complete the tables.

19. Press:	AC/ON	91	STO	9	SUM
Window					
Memory					

20. Press:	AC/ON	91	STO	3	SUM	SUM	SUM
Window							
Memory							

21. Press:	AC/ON	91	STO	3	SUM	CE/C	EXC
Window							
Memory							

22. Press:	AC/ON	91	STO	3	SUM	AC/ON	6	EXC	RCL
Window									
Memory									

★ 23. Press:	AC/ON	91	STO	3	SUM	SUM	RCL	SUM	RCL
Window									
Memory									

Maneuvers with Circles © David A. Page

Remember That Number 63

Try It Out!

1. Use your calculator's memory to calculate the circumference and area of the circle in the sketch at the right.

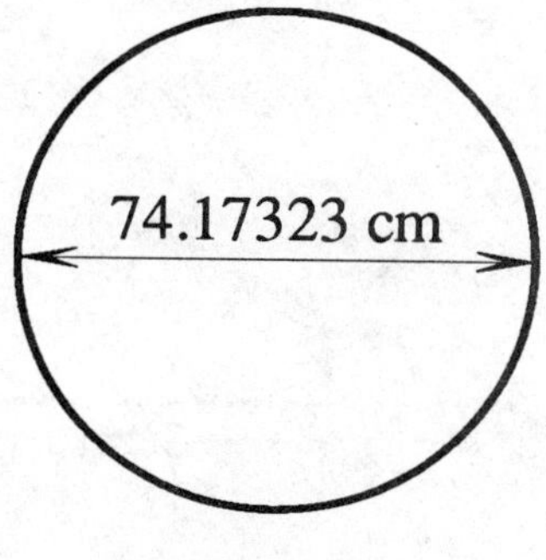

   74.17323 cm

   Figure G

   1a. Complete the following keystrokes to calculate the circumference of the circle.

   | 74.17323 | STO | ☐ | ☐ | ☐ | ☐ |

   Circumference ☐☐☐.☐☐☐☐ cm

   1b. The diameter of the circle is in the memory. [RCL] (Recall) the diameter to calculate the area of the circle. List your keystrokes.

   | RCL | ☐ | ☐ | ☐ | ☐ | ☐ | ☐ | ☐ | ☐ |

   Area _____ cm²
   You'll know.

2. Estimate the area and circumference of the circle below.

   2a. Estimated area _____ cm²

   2b. Estimated circumference _____ cm

3. The circle at the right has a 3.7424103 cm radius. Use the calculator's memory to calculate the area and circumference of the circle. Then compare your answers to your estimates. List your keystrokes.

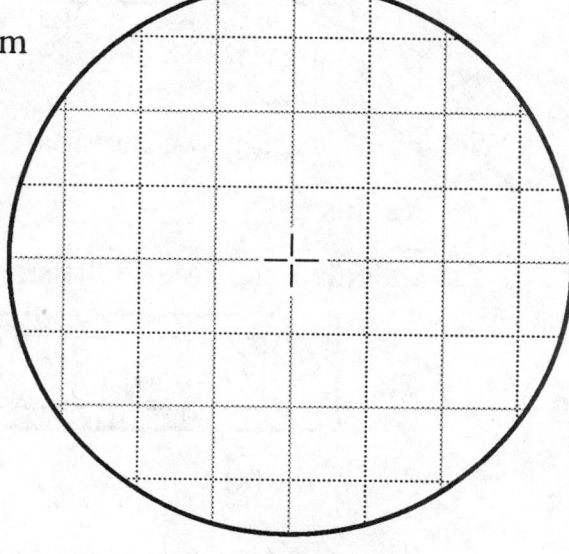

   3a. Area _____ cm²
   You'll know.

   3b. Circumference _____ cm
   Copy window.

   Figure H

© David A. Page   Maneuvers with Circles

64 Chapter 5

4. Two quarter-circles with different radii are put together in the following figure. Use the calculator's memory to calculate the area of the figure.

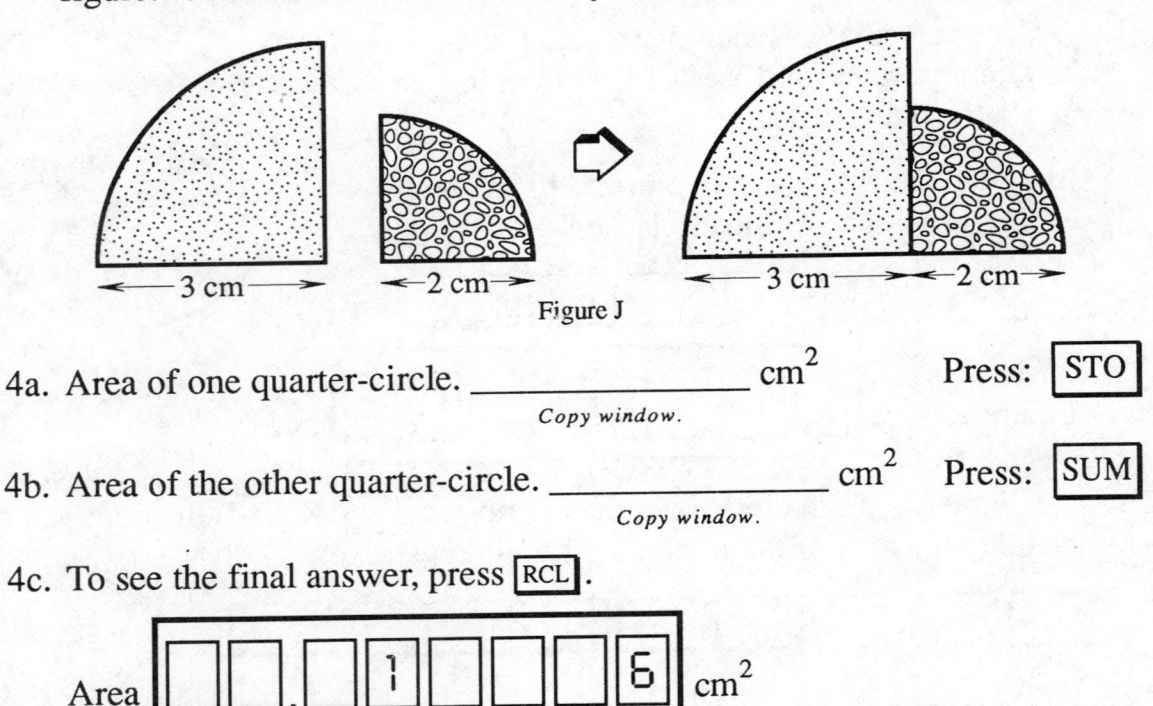

Figure J

4a. Area of one quarter-circle. _____ cm²   Press: STO
 Copy window.

4b. Area of the other quarter-circle. _____ cm²   Press: SUM
 Copy window.

4c. To see the final answer, press RCL.

Area ☐☐.☐1☐☐☐6 cm²

5. Two quarter-circles are put together in the sketch below. Use the following steps to calculate the area of the entire figure.

5a. Calculate the area of the larger quarter-circle.

Area _____ cm²
 Copy window.

Press: STO

5b. Find the radius of the smaller quarter-circle.

Radius _____ cm

5c. Calculate the area of the smaller quarter-circle.

Area ☐8.☐☐3☐☐ cm²

Press: SUM

Figure K

5d. Press RCL to find the total area. _____ cm²
 R to the nearest tenth.

Maneuvers with Circles   © David A. Page

Remember That Number 65

6. The figure at the right is built from four quarter-circles of different sizes. Count square centimeters to estimate the area of the entire shaded figure.

   Estimate _____ cm²

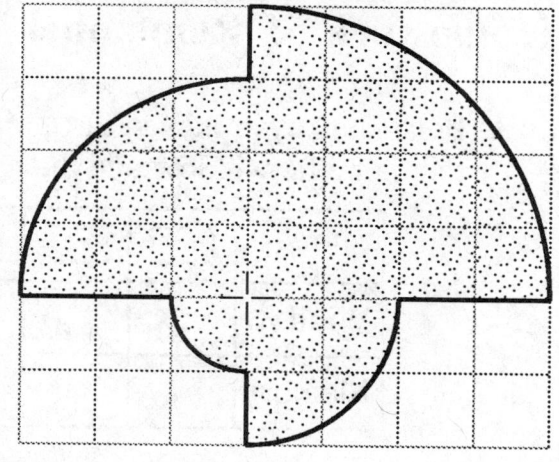

   Figure L

7. Using the steps below as a guide, calculate the area of the shaded quarter-circles. After each quarter-circle, fill in the missing calculator memory keystroke.

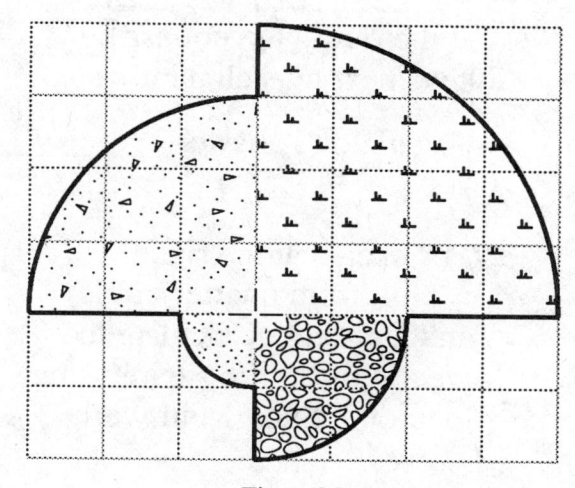

   Figure M

   7a. Area of ⋯ shading _____ cm² Press: [ ]
   Copy window.

   7b. Area of ⋯ shading _____ cm² Press: [ ]
   Copy window.

   7c. Area of ⋯ shading _____ cm² Press: [ ]
   Copy window.

   7d. Area of ⋯ shading _____ cm² Press: [ ]
   Copy window.

   7e. Now find the total area of the shaded figure. Press: [ ] or [ ]

   Area [ ][3][.][ ][ ][1][ ][ ][ ] cm²

   Compare with Problem 6.

© David A. Page   *Maneuvers with Circles*

66 Name _____ Date _____ Class _____

Homework 5: Remember That Number

1. Press the keystrokes in the table. For each keystroke, write the number that is in the window and the number that is in memory.

Press:	AC/ON	74.63	STO	48.37	SUM	EXC	RCL
Window							
Memory							

2. Use STO, SUM, and RCL to do the problem at the right. Do not use +. Write the memory keys next to each number.

    Calculator's answer _____
    You'll know.

    ```
      3.6589741
    845.26981
     51.07521
   + 99.996006
    ```

3. The sketch at the right is built from quarter-circles and a square. Calculate the area of the entire figure using memory. Show your work.

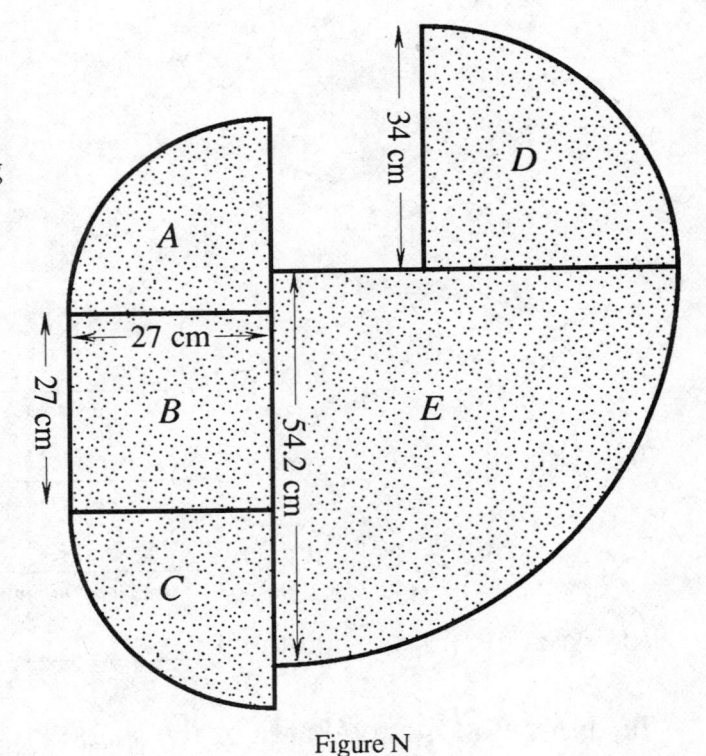

    Area _____ cm²
    R to the nearest hundredth.

Figure N

Maneuvers with Circles © David A. Page

Remember That Number 67

4. Use your calculator's memory to find the area and circumference of the circle at the right.

4a. Area _____ cm²
    R to the nearest tenth.

4b. Circumference _____ cm
    R to the nearest tenth.

4c. List your keystrokes so that you key in the messy radius only once. Finish up with area in the memory and circumference in the window.

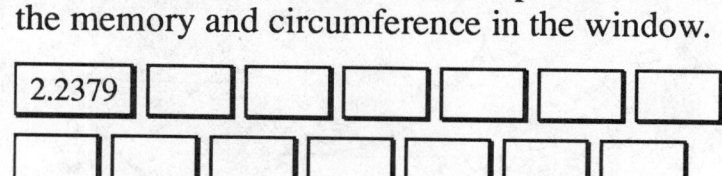

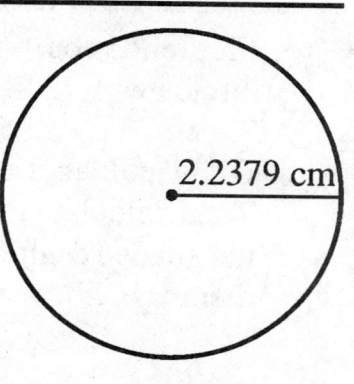

| 2.2379 | | | | | |

| | | | | | |

Figure P — 2.2379 cm

5. The following figure is built from five different circles.

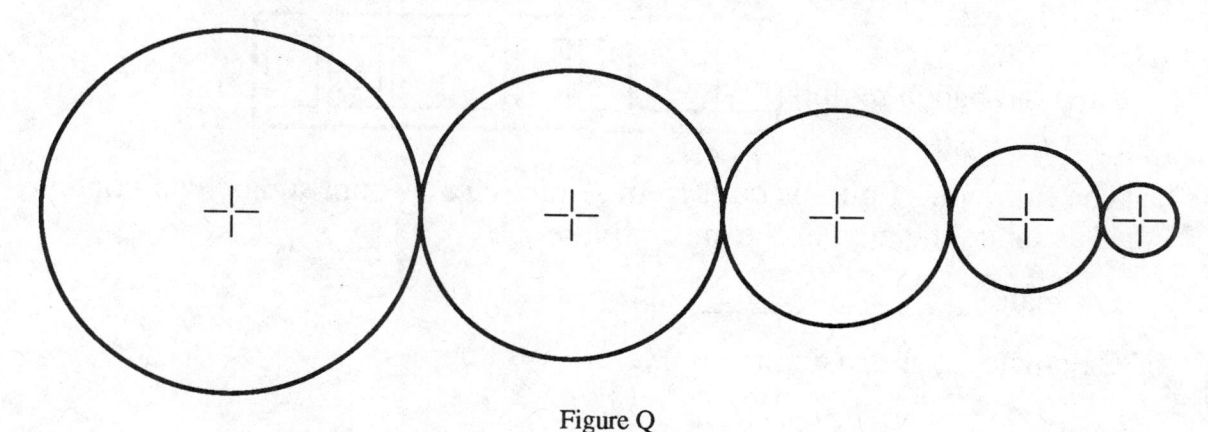

Figure Q

5a. Measure the radius of each circle to the nearest 0.5 cm. Label the radii with your measurements.

5b. Use the calculator's memory to calculate the area of the figure.

Area | | | . | | | 6 | 9 | | cm²

5c. List your keystrokes in one run.

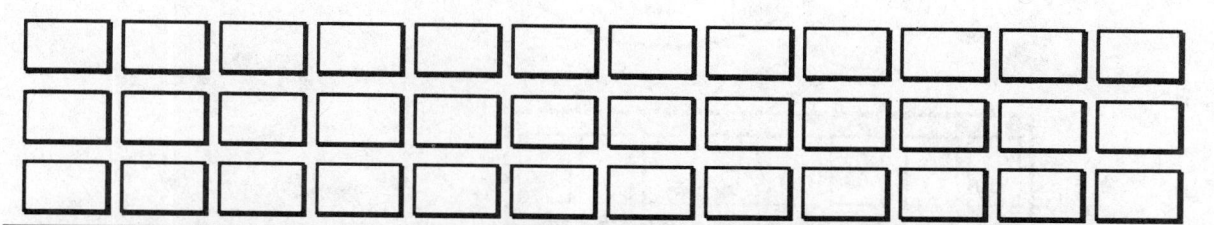

© David A. Page   Maneuvers with Circles

68

6. Figure R is built from quarter-circles with radii of 1.5 cm, 2.5 cm, 3.5 cm, and 4.5 cm. Draw and label each radius. Then calculate the area of the shaded region using memory. Show your work.

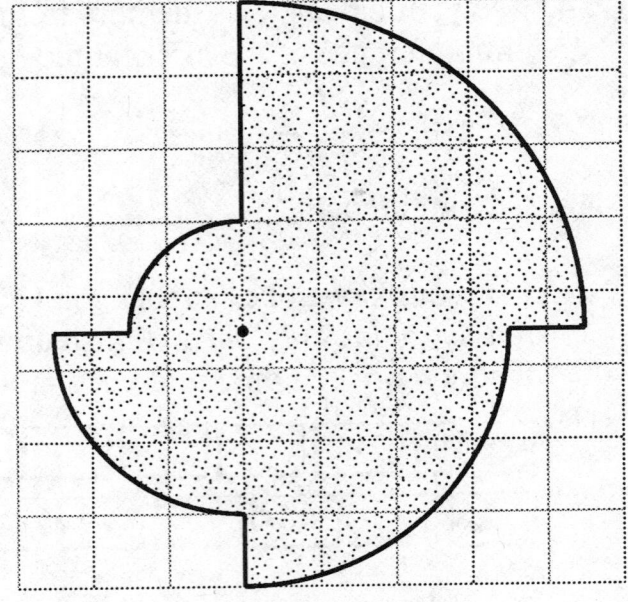

Figure R

Area of shaded region ☐|2|.|☐|☐|☐|2|☐| cm²

7a. The following figure is built from semicircles. Count square centimeters to estimate the area of the entire figure.

Estimate _____ cm²

7b. Calculate its area using memory.

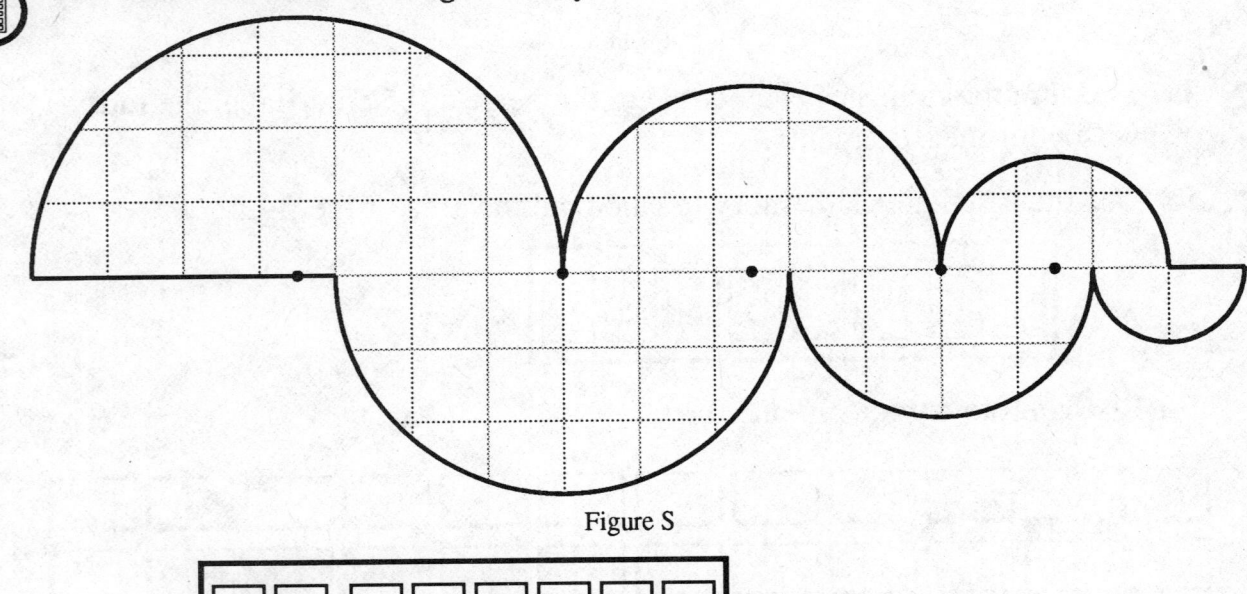

Figure S

Area ☐|☐|.|5|☐|5|☐|☐| cm²

Maneuvers with Circles   © *David A. Page*

Name _____ Date _____ Class _____ 69

6. Circles, Sectors, and Angles

A circle can be built from *sectors*. The following circles are cut into sectors.

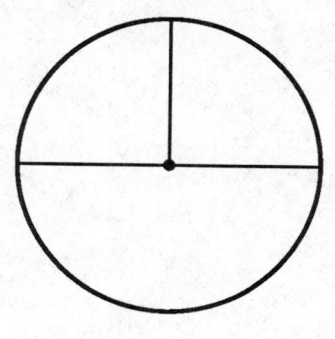

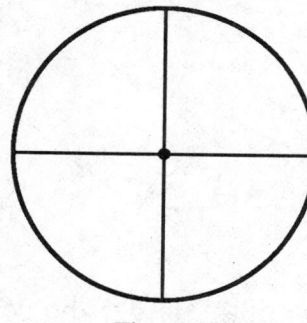

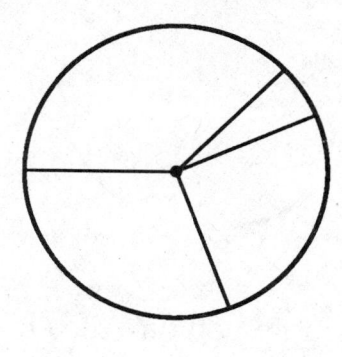

Figure A

Semicircles and quarter-circles are just two examples of sectors.

A sector is built from a curve called an *arc* and two straight sides that meet at the *vertex*.

The straight sides of a sector are radii of the circle. Notice that the vertex of the sector is also the center of the circle.

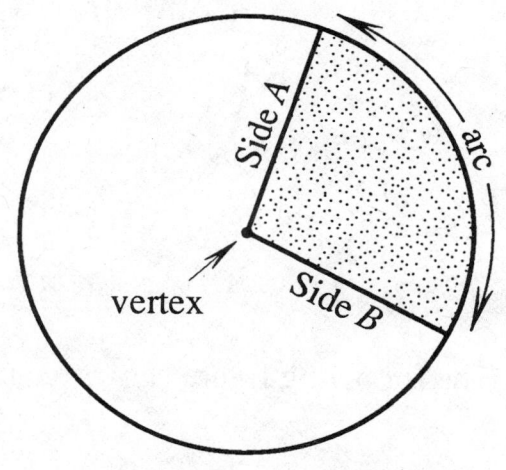

Figure B

1. Look at the shaded part in Figure C.
   Explain why the shaded part is not a sector.

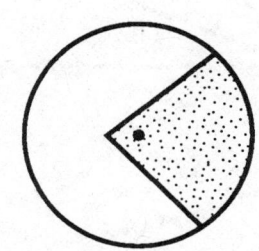

Figure C

© David A. Page   Maneuvers with Circles

Imagine the circle is a clock. The amount of turning from start to finish shows the ***central angle***.

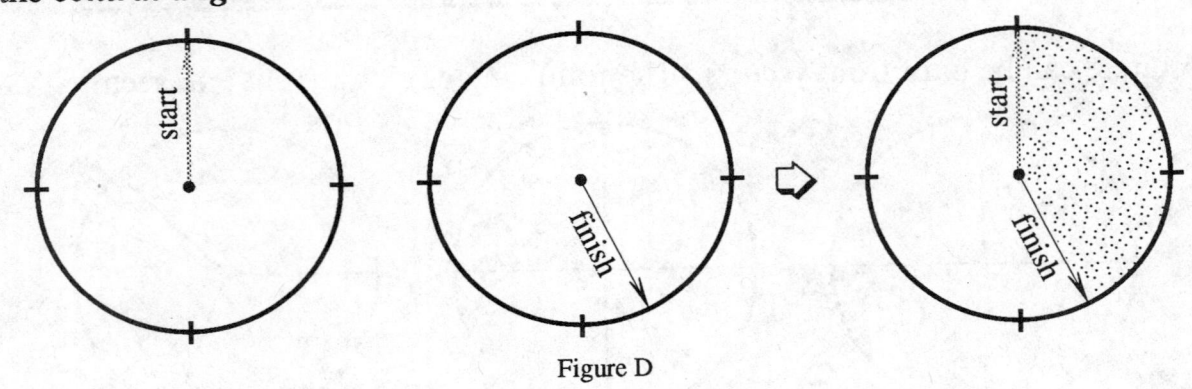

Figure D

The more you turn the clock hand, the larger the central angle becomes.

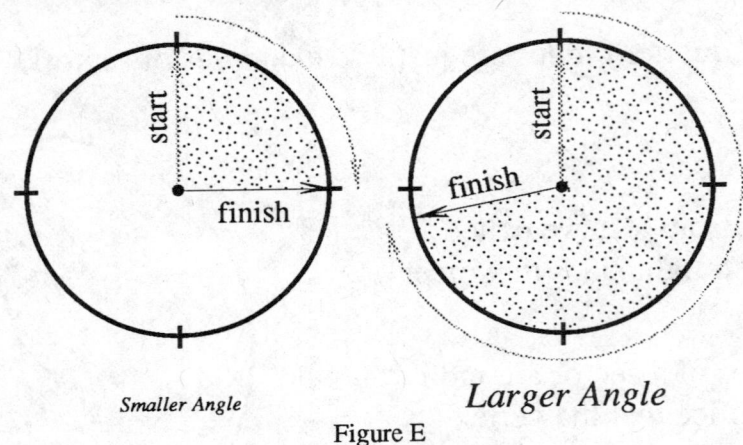

Smaller Angle *Larger Angle*

Figure E

The following figure shows examples of sectors.

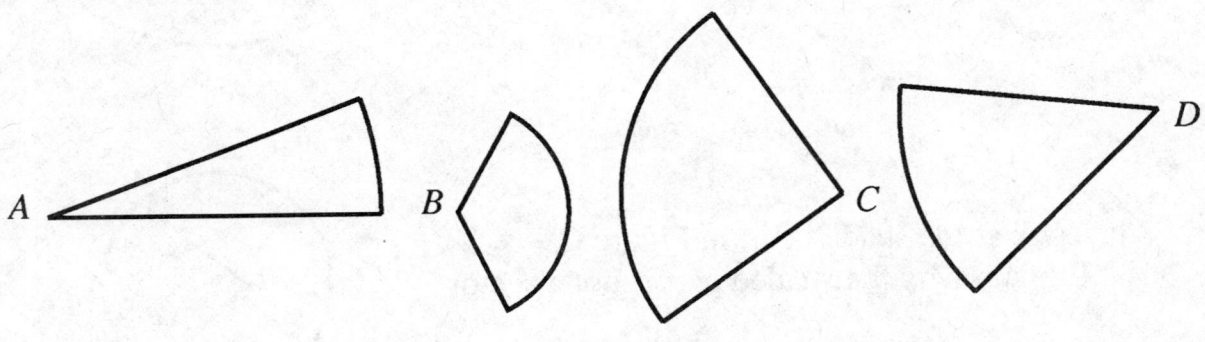

Figure F

2. Compare the amount of turning in each sector above. List the central angles from smallest to largest. Write "*A*," "*B*," "*C*," or "*D*" on the lines.

_____   _____   _____   _____
Smallest   *Largest*

Maneuvers with Circles   © *David A. Page*

Circles, Sectors, and Angles 71

3. Is the following central angle small or large? Remember, think of the amount of turning.

E •————————————————————

Figure G

Answer _____
            Small or Large

We measure central angles in a special unit called *degrees*. The sector above has a small central angle that measures one degree. This can be written as 1°.

Figure H shows three examples of central angles and their measurements.

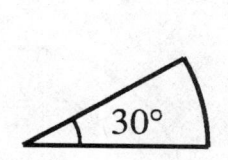

A 30° angle is built from thirty 1° angles.

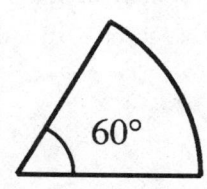

A 60° angle is built from sixty 1° angles.

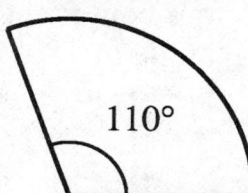

A 110° angle is built from one hundred ten 1° angles.

Figure H

Larger angles have larger degree measures.

You don't always know the degree measure of a central angle. But there are some angles you can easily recognize. A 90° angle is called a *right angle*. Look at the following examples of right angles. Sometimes a "box" is drawn at the vertex to show a 90° angle.

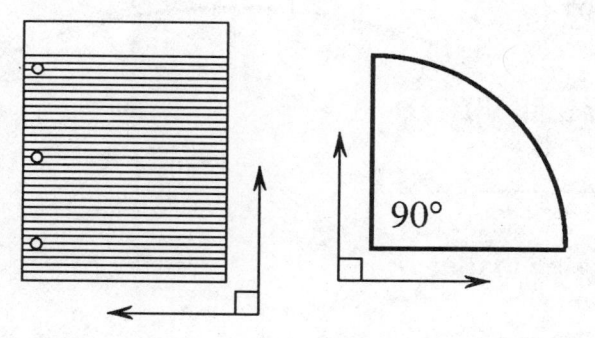

The corner of a sheet of paper is 90°.

A quarter-circle has a central angle of 90° and is built from ninety 1° sectors.

Figure J

© David A. Page   *Maneuvers with Circles*

72   Chapter 6

When two ***congruent*** sectors with a central angle of 90° (two quarter-circles) are put together, a semicircle is formed.

A semicircle has a central angle of 180°.

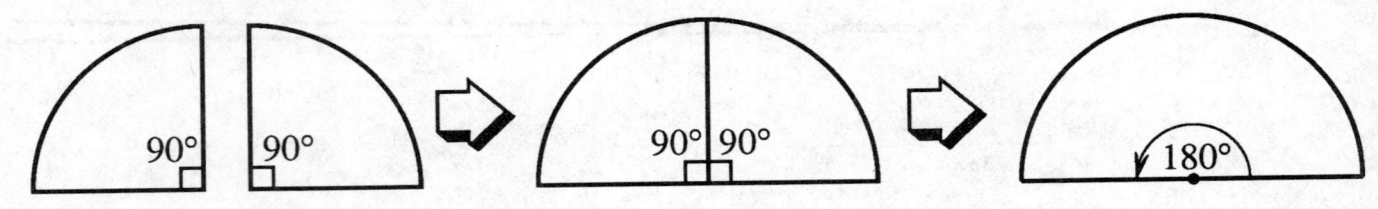

Figure K

When four congruent sectors with central angles of 90° or two sectors with a central angle of 180° are put together, a whole circle is formed.

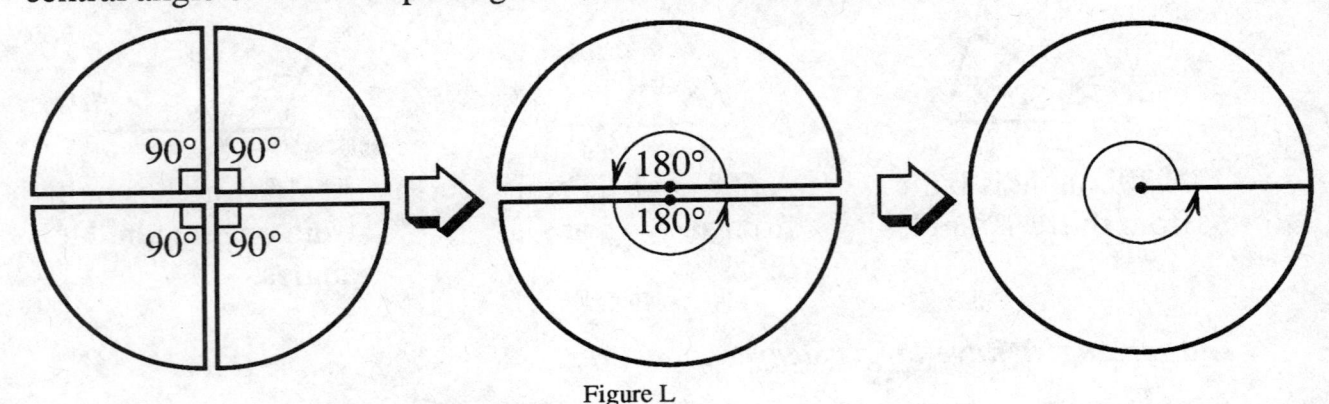

Figure L

4. How many 1° sectors are in a circle? _____

5. Use the following steps to estimate the degree measure of the central angle in Figure M.

5a. Is the central angle smaller or larger than 90°?

Answer _____
      Smaller or Larger

5b. Is the central angle smaller or larger than 180°?

Answer _____
      Smaller or Larger

5c. Estimate the degree measure of the central angle.

Estimate _____°

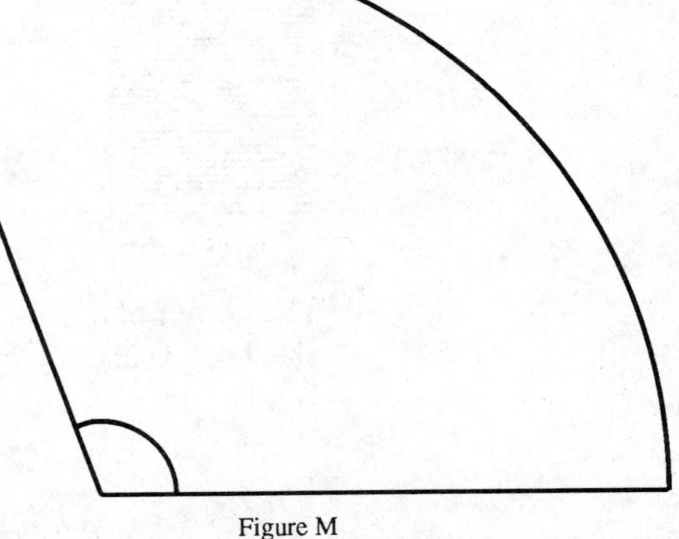

Figure M

Maneuvers with Circles   © David A. Page

Circles, Sectors, and Angles 73

You can find the degree measure of a sector's angle using a ***protractor***.

Chelsea uses Steps **a** through **d** to measure the following central angle.

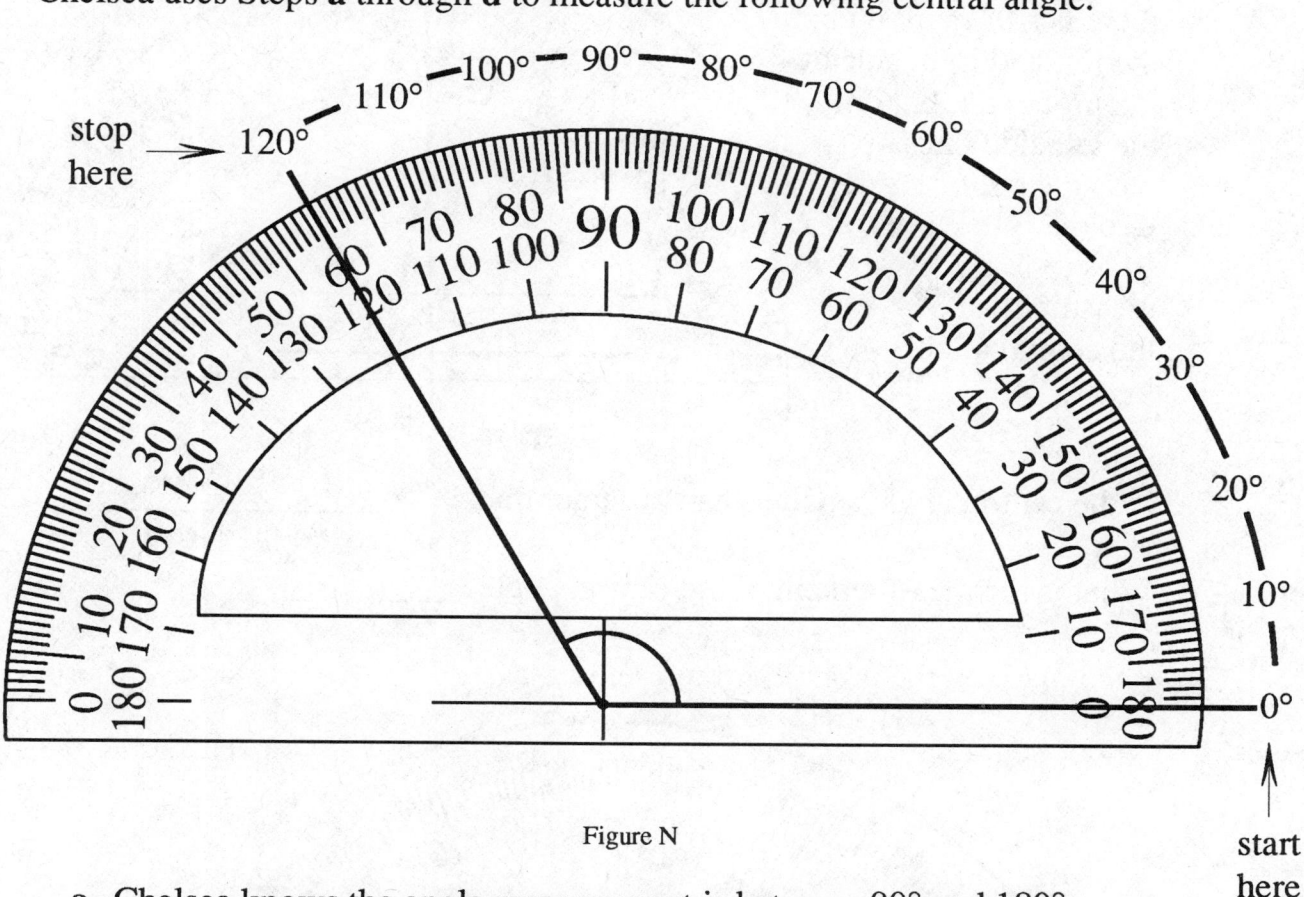

stop here → 120°

Figure N

start here

a. Chelsea knows the angle measurement is between 90° and 180°.

b. Chelsea places the center of the protractor over the vertex of the sector.

c. She lines up one side of the sector with the 0° mark. She begins at 0° and reads the degree marks until she reaches the other side.

d. The side of the sector goes through both the 60° and 120° mark on the protractor. Since the central angle is between 90° and 180°, Chelsea knows that the central angle measures 120°.

6. Use Chelsea's method to measure the central angle in Problem 5.

   Measurement _____°

   Compare with your estimate in Problem 5c.

© David A. Page *Maneuvers with Circles*

74   Chapter 6

7. A protractor is placed on the sector in Figure P. Notice there is a 0° mark on both the left and right side of the protractor, but only the left side lines up with the sector's side. Use the following steps to find the measurement of the central angle.

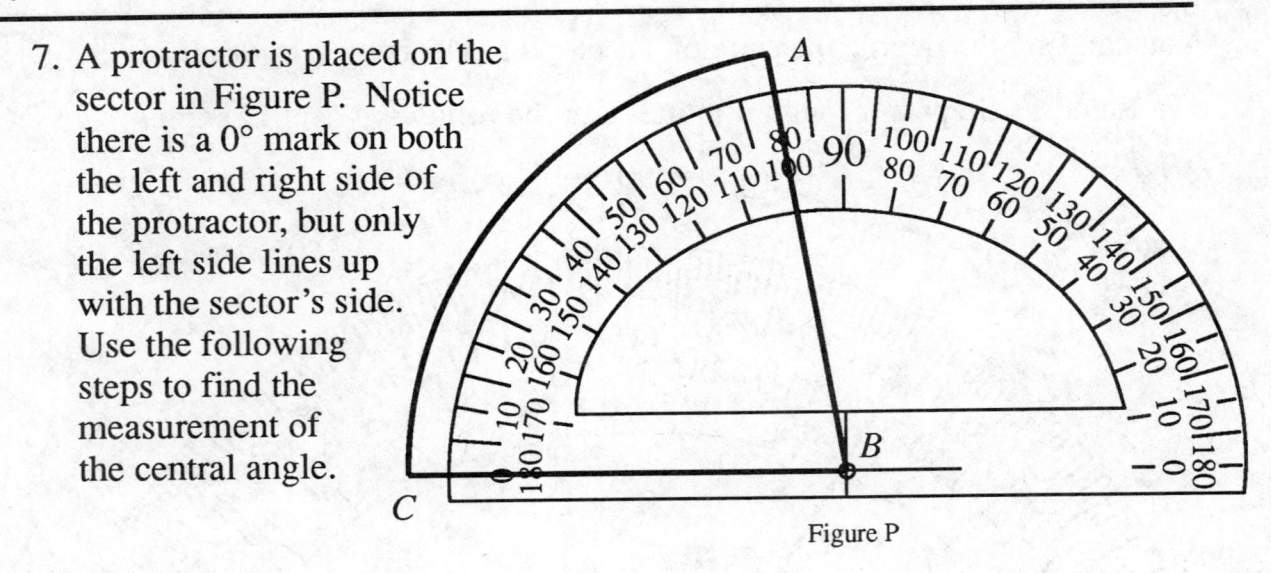

Figure P

7a. Is the central angle smaller than or larger than 90°? _____
    Smaller or Larger

7b. What is the measurement of the central angle? _____°

8. Each mark on the protractor represents 1°.

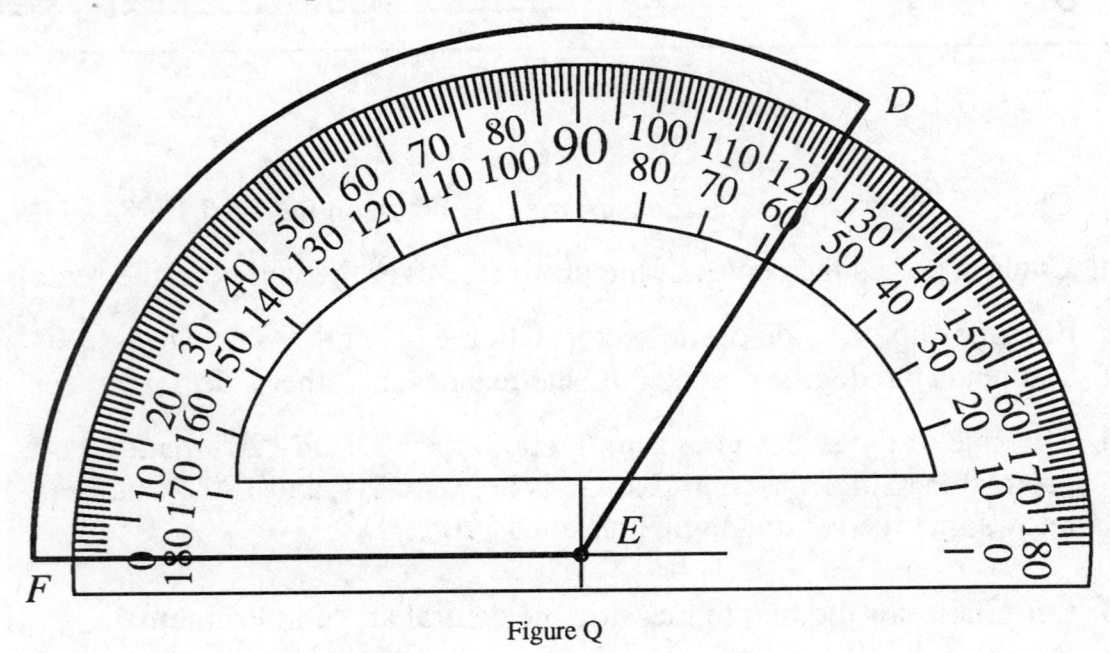

Figure Q

What is the measurement of Sector DEF? _____°

9. Sylvia's answer for Problem 8 was 118°. What did Sylvia do wrong?
_____
_____

Maneuvers with Circles   © David A. Page

Circles, Sectors, and Angles 75

The following figure shows Melanie's paper. She plans to measure the central angle. First she decides the central angle is smaller than 90°.

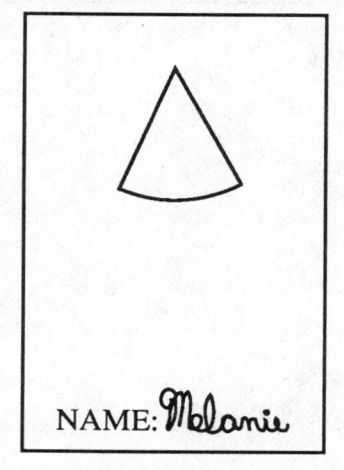

Figure R

To make it easier to measure, she turns the sheet of paper as shown at the right. Melanie places the center of the protractor on the vertex and notices the sides are too short to reach the degree marks on the protractor.

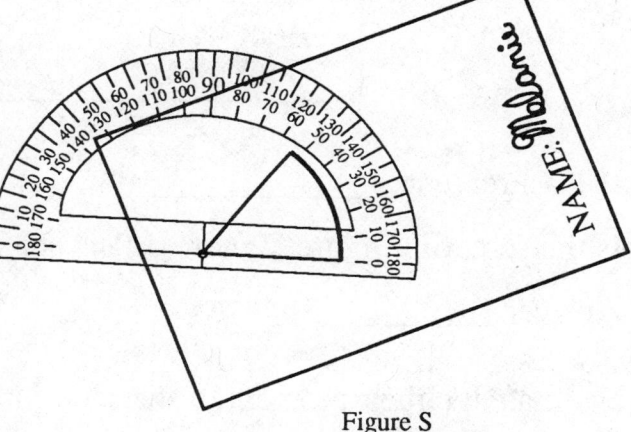

Figure S

Melanie decides to make the sides longer. This is called "*extending the sides*." Once Melanie extends the sides, she can then measure the central angle. Extending the sides does not change the measurement of the central angle.

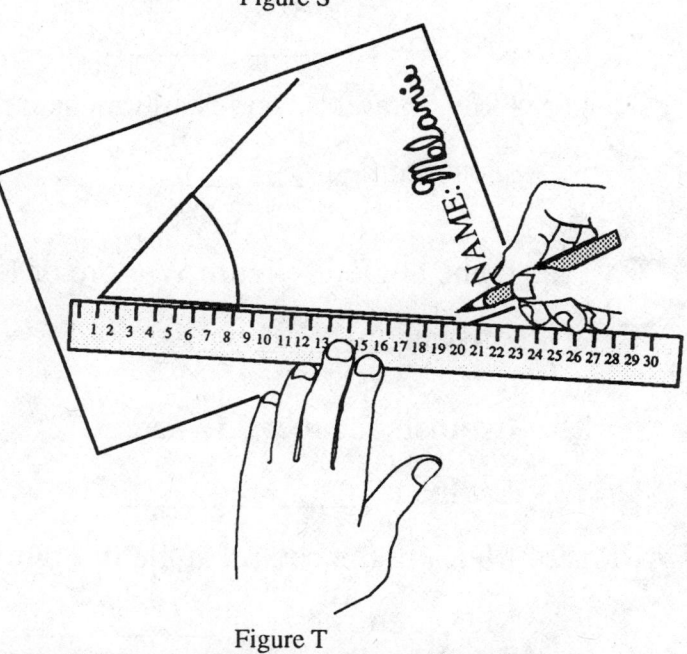

Figure T

© David A. Page   *Maneuvers with Circles*

10. Use Melanie's method to extend the sides in Figure U. Measure the sector's angle.

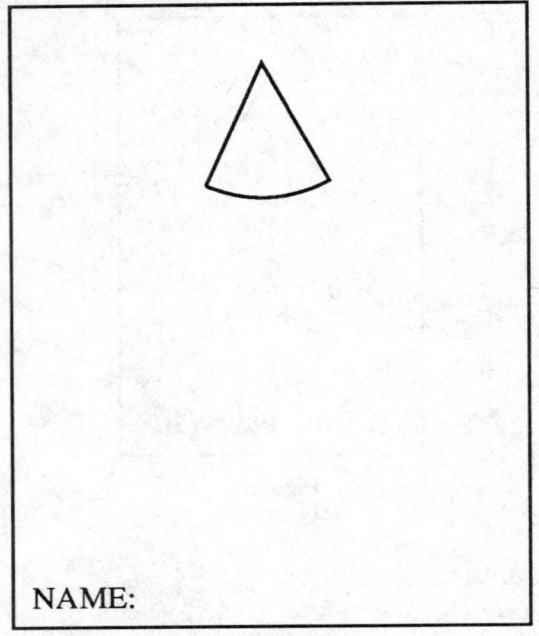

Figure U

Measurement _____°

11a. Is the sector in Figure V more or less than 90°?

Answer _____
      More or Less

11b. Estimate the measurement of the sector in Figure V.

Estimate _____°

Figure V

11c. Now extend the sides and measure the angle.

Measurement _____°
   Compare with your estimate.

12a. Is the angle in Figure W more or less than 90°?

Answer _____
      More or Less

12b. Estimate the measurement.

Estimate _____°

12c. Measure the central angle in Figure W.

Figure W

Measurement _____°

Circles, Sectors, and Angles

Area of Sectors

The sector or quarter-circle at the right has a central angle of 90° and a radius of 3 cm.

Find the area of the 90° sector using the following steps.

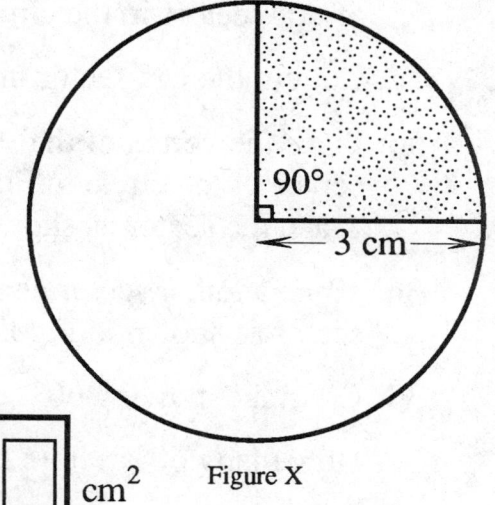

Figure X

1a. Find the area of the whole circle.

Area of circle ☐☐.2☐☐☐3☐ cm²

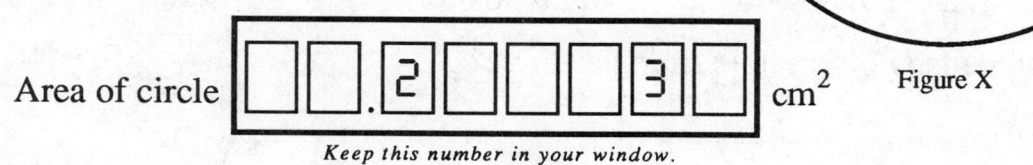

Keep this number in your window.

1b. How many 90° sectors are in a circle? _____

1c. Divide the area of the circle by the number of 90° sectors in the circle to find the area of one sector.

Area of 90° sector ☐.0☐☐8☐☐ cm²

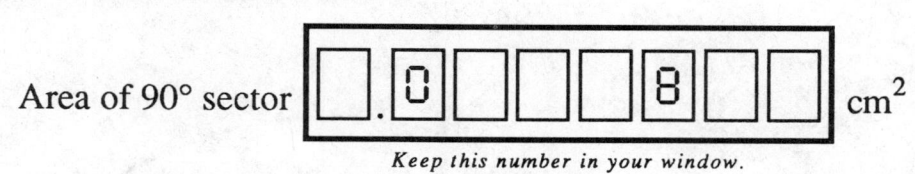

Keep this number in your window.

2. A 270° sector is built from three 90° sectors. Calculate the area of the following 270° sector.

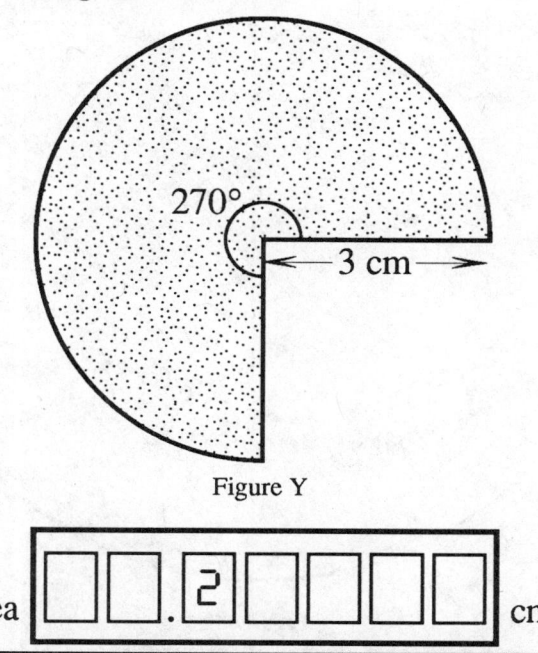

Figure Y

Area ☐☐.2☐☐☐ cm²

© David A. Page   Maneuvers with Circles

Chapter 6

3. Use the following steps to find the number of 60° sectors in the circle below.

3a. Cut out the 60° sector in Figure BB.

3b. Place the vertex of the sector on the center of the circle in Figure CC. Trace the sector as shown at the right.

3c. Trace another sector next to the first sector as shown in Figure AA.

3d. Continue drawing 60° sectors until no more fit.

3e. How many 60° sectors fit in the circle?

Answer _____

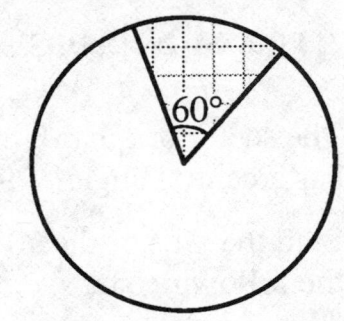

Figure Z

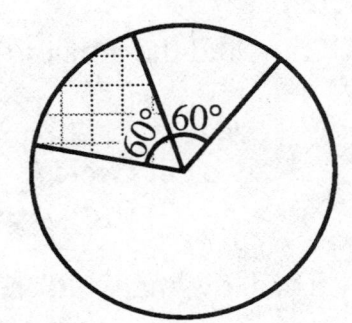

Figure AA

Save your sector.

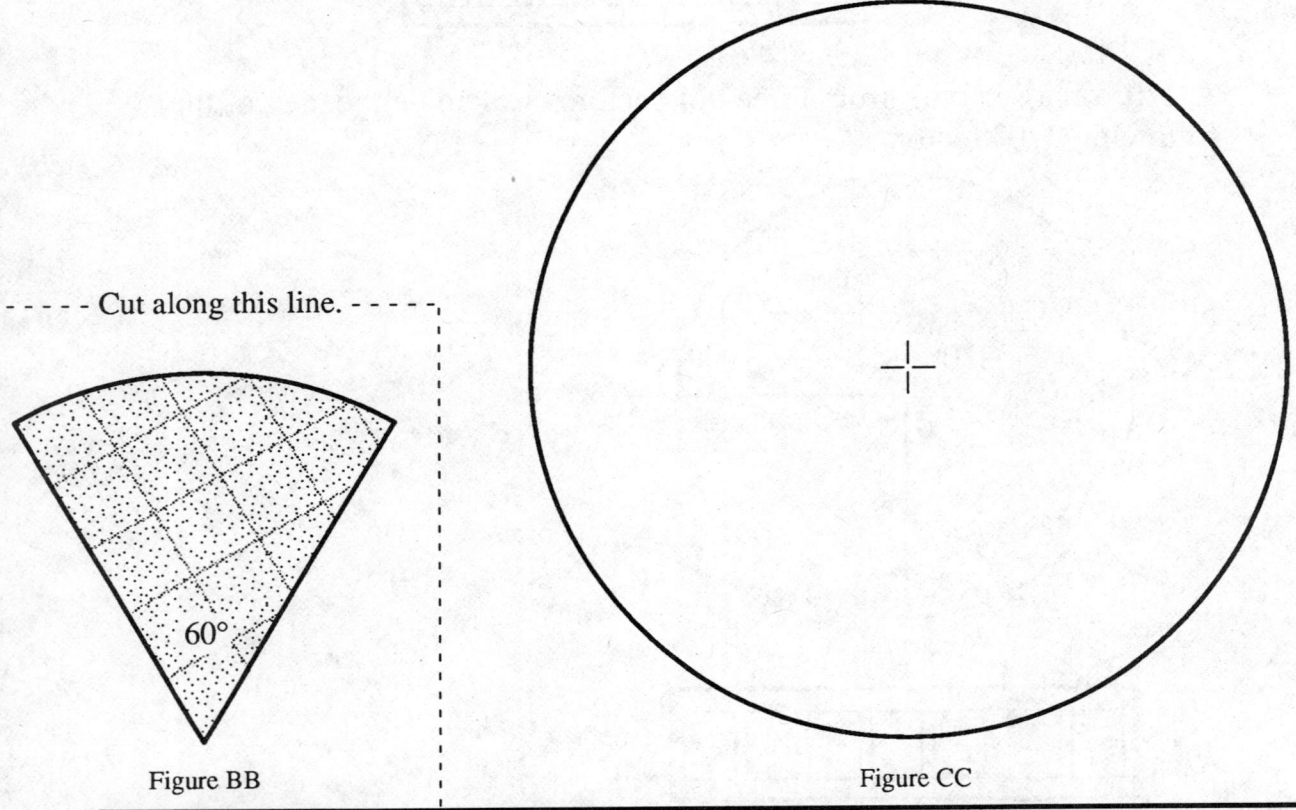

- - - - - Cut along this line. - - - - -

Figure BB

Figure CC

Maneuvers with Circles

© David A. Page

Circles, Sectors, and Angles

4. Glue your 60° sector in the space at the right.

Glue or tape here.

5. Count the square centimeters in the 60° sector above to estimate the area. Circle the best estimate.

   1 cm²   5 cm²   10 cm²   60 cm²

6. Calculate the area of the 60° sector using the following steps.

6a. The radius of the circle in Figure CC is 5 cm. Calculate the circle's area.

   Area of circle (_ 8 . _ _) cm²
   R to the nearest hundredth.

6b. How many 60° sectors fit in the circle? _____

6c. Divide the area of the circle by the number of 60° sectors in the circle to find the area of one 60° sector.

   Area of 60° sector (_ _ . _) cm²
   R to the nearest tenth.

80 Chapter 6

7a. Terry started to trace 45° sectors in the circle below. Draw in the other radii to show the number of 45° sectors in a circle.

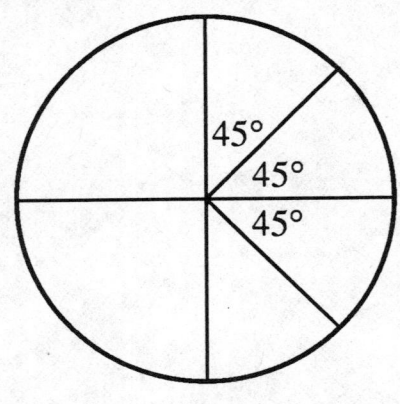

Figure DD

7b. How many 45° sectors build a circle? _____

8a. Joline did the same thing as Terry but with 30° sectors. Finish drawing in radii below to show the number of 30° sectors in a circle.

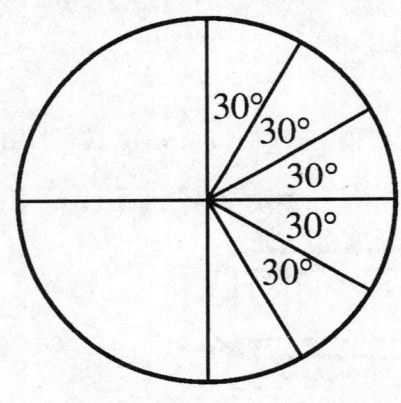

Figure EE

8b. How many 30° sectors build a circle? _____

9. Julie started to draw 12° sectors in the circle at the right. Instead, she used a faster way to find the number of 12° sectors in a circle. Julie knew a circle has 360°, so she divided it into 12° parts.

9a. 360° ÷ 12° = _____

9b. There are _____ 12° sectors in a circle.

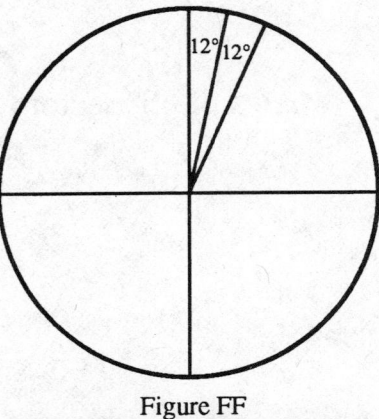

Figure FF

Maneuvers with Circles © David A. Page

Circles, Sectors, and Angles

81

10. The circle in the sketch at the right has a radius of 8 cm. Calculate the area of the circle.

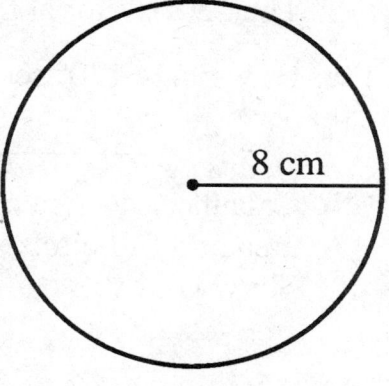

8 cm

Area _____ cm²
R to the nearest hundredth.

Figure GG

11. Complete the table to find the area of each sector with an 8 cm radius.

	Degrees in Whole Circle	Central Angle	Number of Sectors in Circle	Area of Sector with Radius of 8 cm *R to the nearest hundredth.*
a.	360°	60°		201.06 cm² ÷ 6 = 33.51 cm² *Area of Whole Number of Sectors Area of One*
b.	360°	45°		201.06 cm² ÷ ___ = _____
c.		30°		
d.		15°		
e.		12°		
f.		11°		*Copy window.*
g.		10°		

12. The radius of each sector in the table is 8 cm. When the radius is the same, what happens to the area of the sector as the central angle gets smaller?

_____

© David A. Page

Maneuvers with Circles

13. The 120° sector at the right has a radius of 3 cm.

13a. How many 120° sectors are there in a circle?

Answer _____

13b. Calculate the area of the sector. Remember, find the area of a full circle with a radius of 3 cm first. Show your work.

Figure HH

Area of 120° sector _____ cm²

R to the nearest tenth.

14. The following 1° sector has a radius of 11 cm. Calculate its area using the steps below.

Figure JJ

14a. Calculate the area of a circle with a radius of 11 cm.

Area ⬚⬚⬚.1⬚3⬚⬚ cm²

14b. How many 1° sectors build a circle? _____

14c. Calculate the area of the 1° sector. ⬚.⬚⬚ cm²

R to the nearest hundredth.

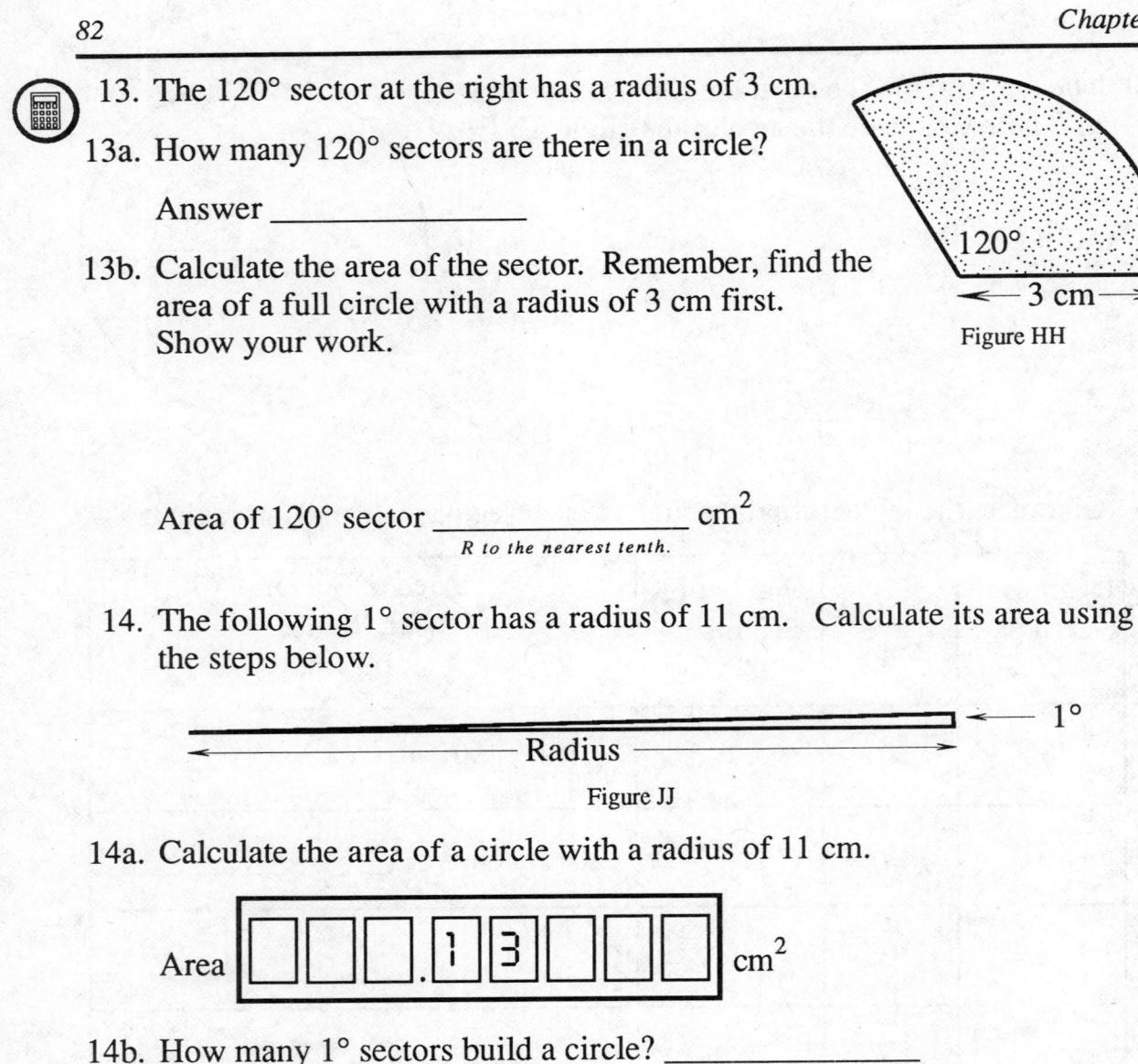

15. Five of the 1° sectors from Problem 14 are put together to make the following 5° sector. Calculate the area of the 5° sector.

Figure KK

Area of 5° sector _____ cm²

R to the nearest tenth.

Maneuvers with Circles © *David A. Page*

Circles, Sectors, and Angles

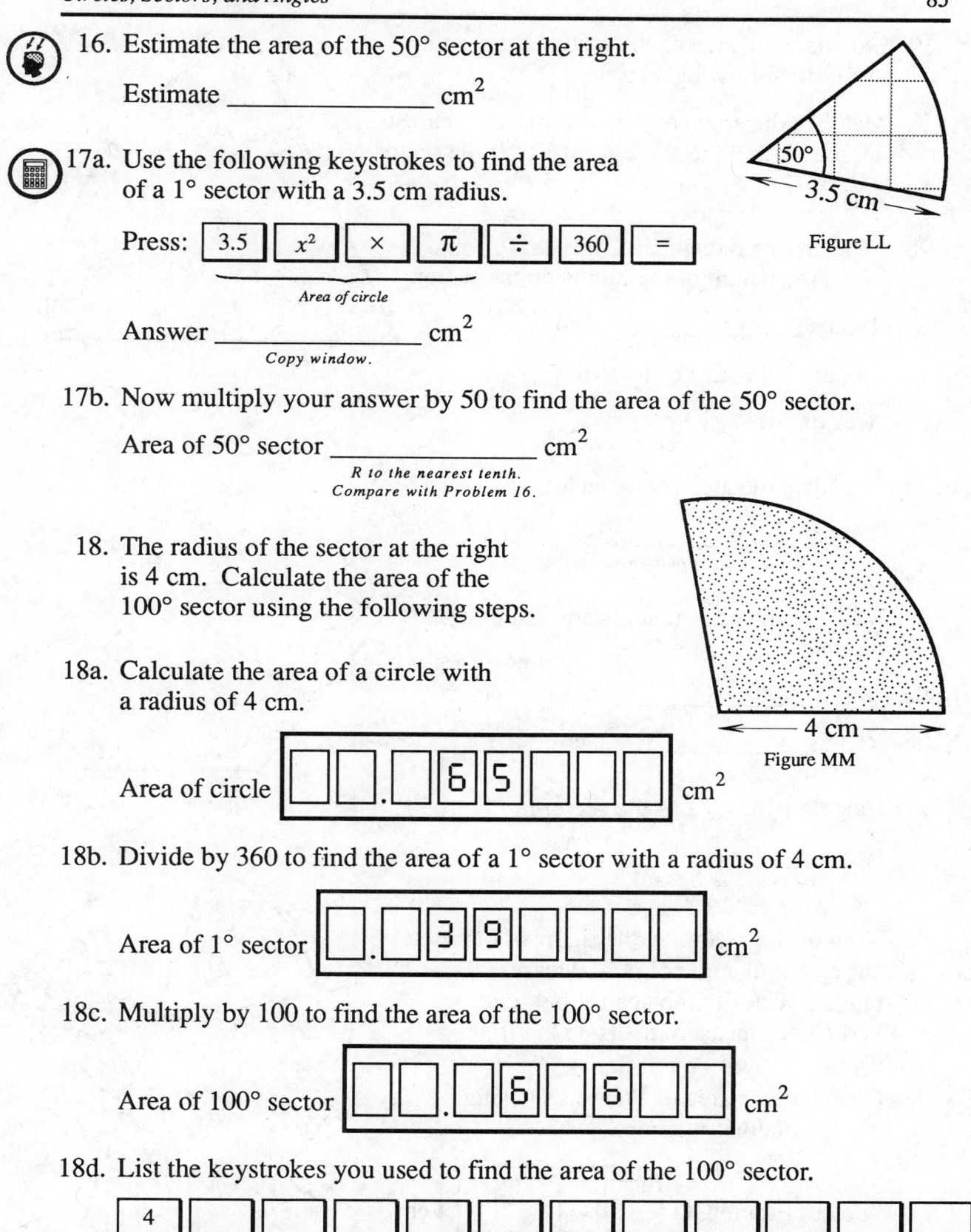

16. Estimate the area of the 50° sector at the right.

   Estimate _____ cm²

17a. Use the following keystrokes to find the area of a 1° sector with a 3.5 cm radius.

   Press: | 3.5 | x^2 | × | π | ÷ | 360 | = |

   Area of circle

   Answer _____ cm²
   Copy window.

Figure LL

17b. Now multiply your answer by 50 to find the area of the 50° sector.

   Area of 50° sector _____ cm²
   R to the nearest tenth.
   Compare with Problem 16.

18. The radius of the sector at the right is 4 cm. Calculate the area of the 100° sector using the following steps.

18a. Calculate the area of a circle with a radius of 4 cm.

   Area of circle | ☐ | ☐ | . | ☐ | 6 | 5 | ☐ | ☐ | cm²

Figure MM

18b. Divide by 360 to find the area of a 1° sector with a radius of 4 cm.

   Area of 1° sector | ☐ | . | ☐ | 3 | 9 | ☐ | ☐ | ☐ | cm²

18c. Multiply by 100 to find the area of the 100° sector.

   Area of 100° sector | ☐ | ☐ | . | ☐ | 6 | 6 | ☐ | ☐ | cm²

18d. List the keystrokes you used to find the area of the 100° sector.

   | 4 | ☐ | ☐ | ☐ | ☐ | ☐ | ☐ | ☐ | ☐ | ☐ | ☐ | ☐ |
   You do not need to use all the keystroke boxes.

© David A. Page   Maneuvers with Circles

84   Chapter 6

19. Calculate the area of the sector at the right using the following steps.

19a. Measure the angle of the sector to the nearest degree. Label the angle measure on the sector.

  Angle _____ °

19b. Measure the radius to the nearest 0.1 cm. Label the length of the radius on the sector.

  Radius _____ cm

19c. Calculate the area of the circle.

  Area of circle _____ cm^2
  _{Copy window.}

19d. Calculate the area of the sector.

  Area of sector _____ cm^2
  _{R to the nearest tenth.}

Figure NN

20. Measure the central angle and radius of the sector in Figure PP.

20a. Angle _____ °

20b. Radius _____ cm

21. Calculate the area of the sector in Figure PP.

  Area _____ cm^2
  _{R to the nearest tenth.}

Figure PP

22. Each of the sectors in the figure at the right has a central angle of 60°. Measure the radius of each sector to the nearest half cm. Write your measurements on the figure.

23. Calculate the area of Figure QQ using the calculator's memory.

  Area of Figure QQ ☐☐.☐☐ cm^2
  _{R to the nearest hundredth.}

Figure QQ

Maneuvers with Circles   © *David A. Page*

Name _____ Date _____ Class _____ 85

Homework 6: Circles, Sectors, and Angles

1. Write "could be" or "crazy" after each statement.

   ∠A is a right angle. _____

   ∠A is 175°. _____

   ∠A is smaller than ∠B. _____

   ∠B is smaller than 180°. _____

   ∠B is larger than 90°. _____

   ∠B is 10°. _____

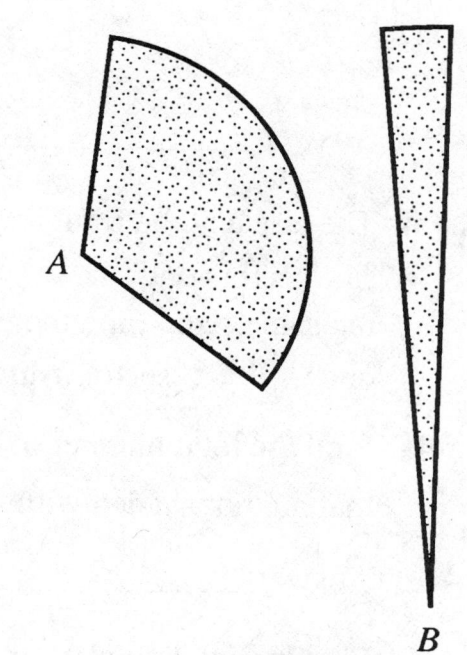

Figure RR

2a. Look at the angles in the figure below. Write an "S" on the sector with the smallest central angle. Write an "L" on the sector with the largest central angle.

2b. Measure each central angle. Don't forget to extend the sides. Write the measurement next to each angle.

Figure SS

© David A. Page

Maneuvers with Circles

3. Find the number of degrees in the sector with the ⟨rocky⟩ shading at the right. Show your work.

   Answer _____°

   Figure TT

4. The circle at the right is cut into five sectors. The sectors with the ⟨dotted⟩ shading are the same size. Find the number of degrees in one ⟨dotted⟩ sector using the following steps.

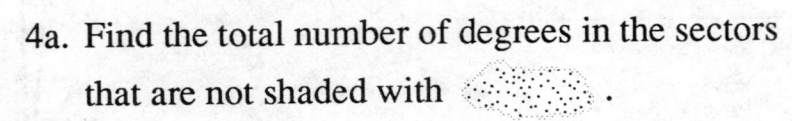

   4a. Find the total number of degrees in the sectors that are not shaded with ⟨dotted⟩.

   Answer _____°

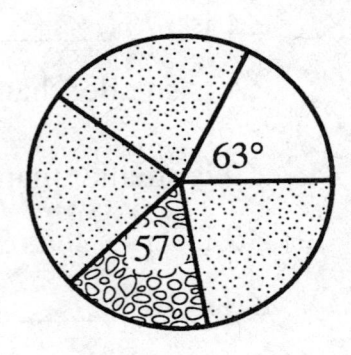

   Figure UU

   4b. Find the total number of degrees in the three ⟨dotted⟩ sectors.

   Answer _____°

   4c. Find the number of degrees in one ⟨dotted⟩ sector.

   Answer _____°

5. Find the number of 72° sectors that fit in a circle. Show your work.

   Answer _____

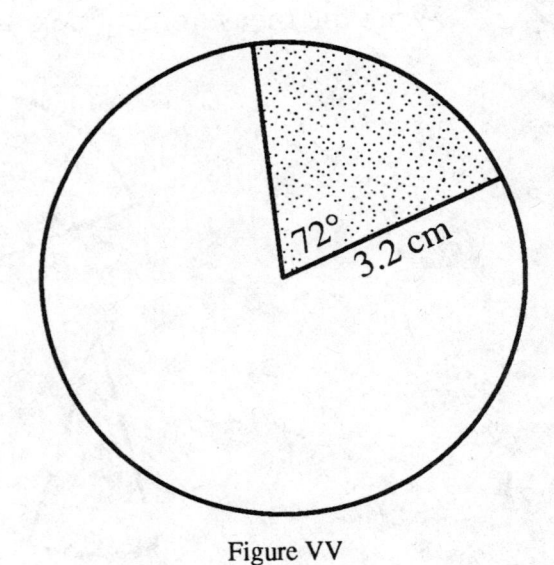

6. Calculate the area of the 72° sector in Figure VV. Show your work.

   Figure VV

   Area of 72° sector _____ cm^2
   R to the nearest tenth.

Circles, Sectors, and Angles   87

7. Calculate the area of the 18° sector in Figure WW. Show your work.

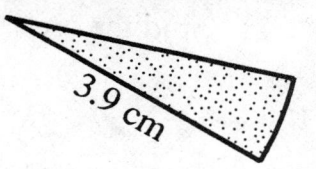

Figure WW (3.9 cm)

Area of sector _____ cm²
R to the nearest tenth.

8. A sector has a central angle of 32° and a radius of 4.4 cm. Calculate the area of this 32° sector.

Area of sector _____ cm²
R to the nearest tenth.

9a. Calculate the area of the 300° sector in the sketch at the right.

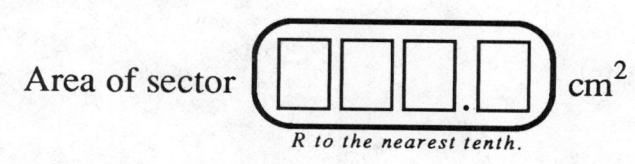

Area of sector (☐ ☐ ☐ . ☐) cm²
R to the nearest tenth.

9b. List the keystrokes you used to calculate the area of the sector.

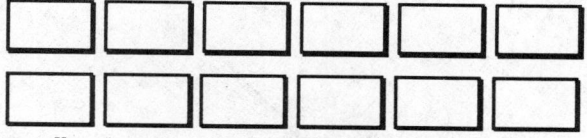

☐ ☐ ☐ ☐ ☐ ☐
☐ ☐ ☐ ☐ ☐
You do not need to use all the keystroke boxes.

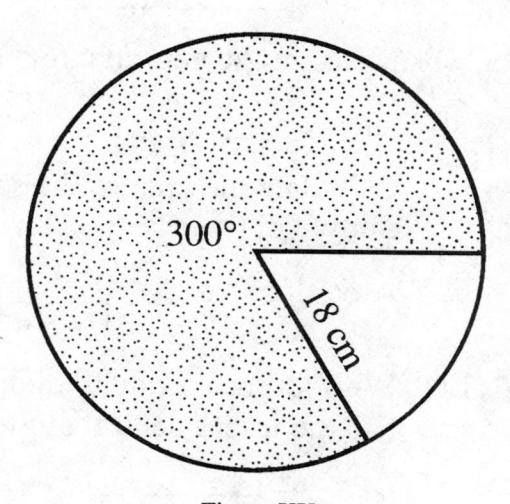

Figure XX (300°, 18 cm)

10. Twelve of the sectors at the right fit in a circle.

10a. Without measuring, what is the degree measure of the sector?

Answer _____ °

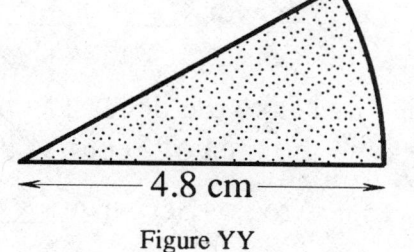

10b. Calculate the area of the sector. Show your work.

Figure YY (4.8 cm)

Area of sector _____ cm²
R to the nearest tenth.

© David A. Page   *Maneuvers with Circles*

11. Find the area of the following sector using the steps below.

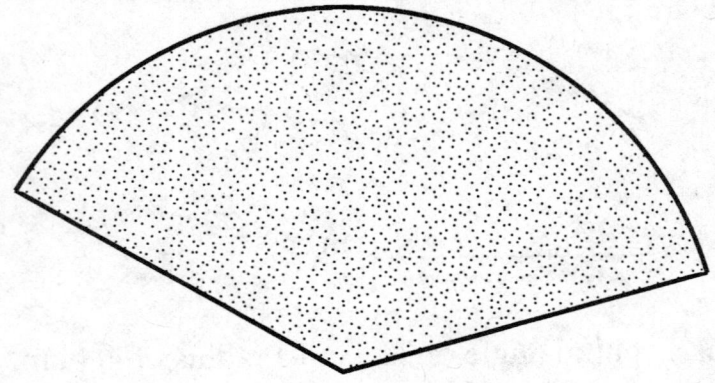

Figure ZZ

11a. Measure the radius of the circle to the nearest 0.1 cm.

Radius _____ cm

11b. Measure the central angle of the sector to the nearest degree.

Answer _____ °

11c. Calculate the area of the sector.
Show your work.

Area of sector _____ cm^2
R to the nearest tenth.

12a. Without measuring, which sector at the right has the larger central angle, A or B?

Answer _____
A or B

12b. Predict which sector has the larger area.

Answer _____
A or B

13. Make the necessary measurements and find the area of Sector A and Sector B.

13a. Area of Sector A _____ cm^2
Copy window.

13b. Area of Sector B _____ cm^2
Copy window.

13c. Were your answers to Problems 12a and 12b correct?

Answer _____
Yes or No

Figure AAA

Maneuvers with Circles © David A. Page

Name _____ Date _____ Class _____ 89

7. Circles with Holes

1. Draw a circle with a radius of 2 cm anywhere inside the circle below.

2a. Cut out the large circle at the bottom of the page.

2b. Fold it in half to find a diameter of the circle. Measure the diameter to the nearest whole cm.

   Diameter _____ cm

2c. How long is the radius? _____ cm

2d. Calculate the area of the large circle.

   Area of large circle _____ cm²
   R to the nearest tenth.

3. Calculate the area of the smaller circle.

   Area of smaller circle _____ cm²
   R to the nearest tenth.

4. Cut out the smaller circle. You will have a circle with a hole in it.

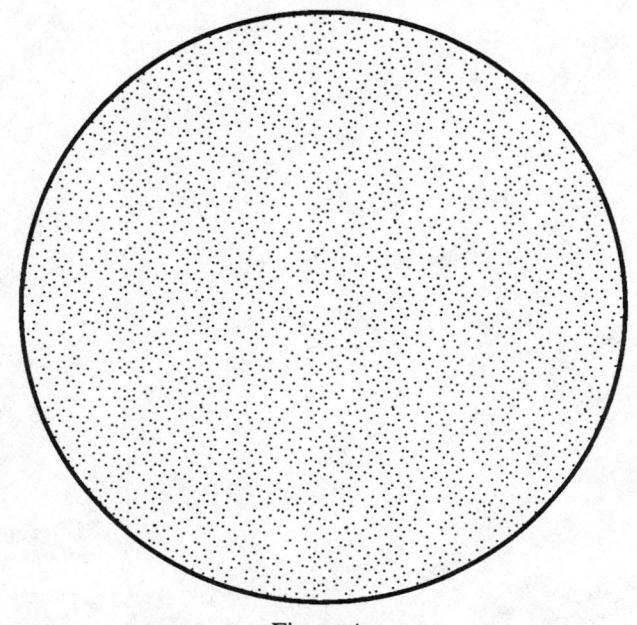

- - - - - - - - - - - - - - - - - - - Cut below this line. -

Figure A

© David A. Page

Maneuvers with Circles

5. Glue or tape the circle with
   the hole at the right. This is
   the part left over.

                                            Glue or tape here.

6. Find the area of the part left
   over using the following steps.

   Area of large circle − Area of small circle = Area of circle with hole

   _____ − _____ = _____ cm^2
   <small>Copy answer from</small>   <small>Copy answer from</small>
   <small>Problem 2d.</small>   <small>Problem 3.</small>

7. Did everyone draw the small circle in the same place? _____
   <small>Yes or No</small>

8a. When the small circle is in a different place, is the area left over the same?

    Answer _____
    <small>Yes or No</small>

8b. Why or why not? _____

Maneuvers with Circles   © *David A. Page*

Circles with Holes

9. Use the centimeter grid to estimate the area of the shaded part in Figure B.

   Estimate _____ cm²

10. Calculate the area of the shaded part in Figure B using the following steps.

   10a. How long is the radius of the larger circle?

   Radius of larger circle _____
   Put in units.

   10b. Calculate the area of the larger circle.

   Area of larger circle _____ cm²
   R to the nearest hundredth.

   10c. How long is the radius of the white circle? _____ cm

   10d. What is the area of the white circle? _____ cm²
   R to the nearest hundredth.

   10e. Subtract to find the area of the shaded part.

   Area of shaded part _____ cm²
   R to the nearest tenth.
   Compare with your estimate in Problem 9.

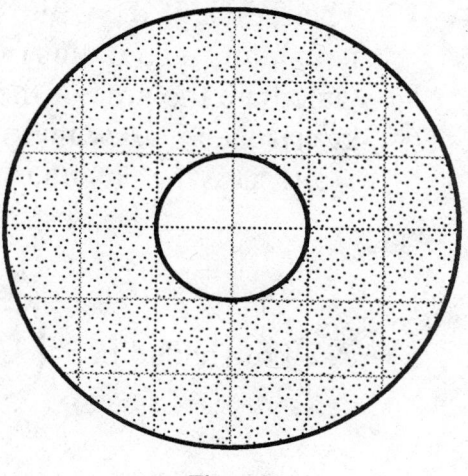

Figure B

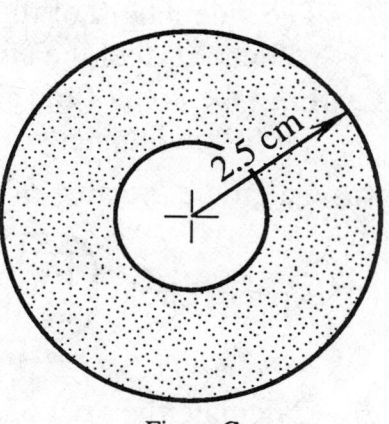

11a. Draw a radius for the white circle in Figure C. Measure and label the radius to the nearest 0.1 cm.

11b. Calculate the area of the white circle.

   Area of white circle _____ cm²
   Copy window.

11c. Calculate the area of the larger circle.

   Area of larger circle _____ cm²
   Copy window.

11d. Calculate the area of the shaded part.

   Area of shaded part _____ cm²
   R to the nearest tenth.

Figure C (2.5 cm)

© David A. Page

Maneuvers with Circles

12. The radius of the white circle in the sketch at the right is 2.3 cm. The radius of the large circle is 9.45 cm. Calculate the area of the shaded part. List your keystrokes. Draw in your own keystroke boxes. Hint: Use the calculator's memory.

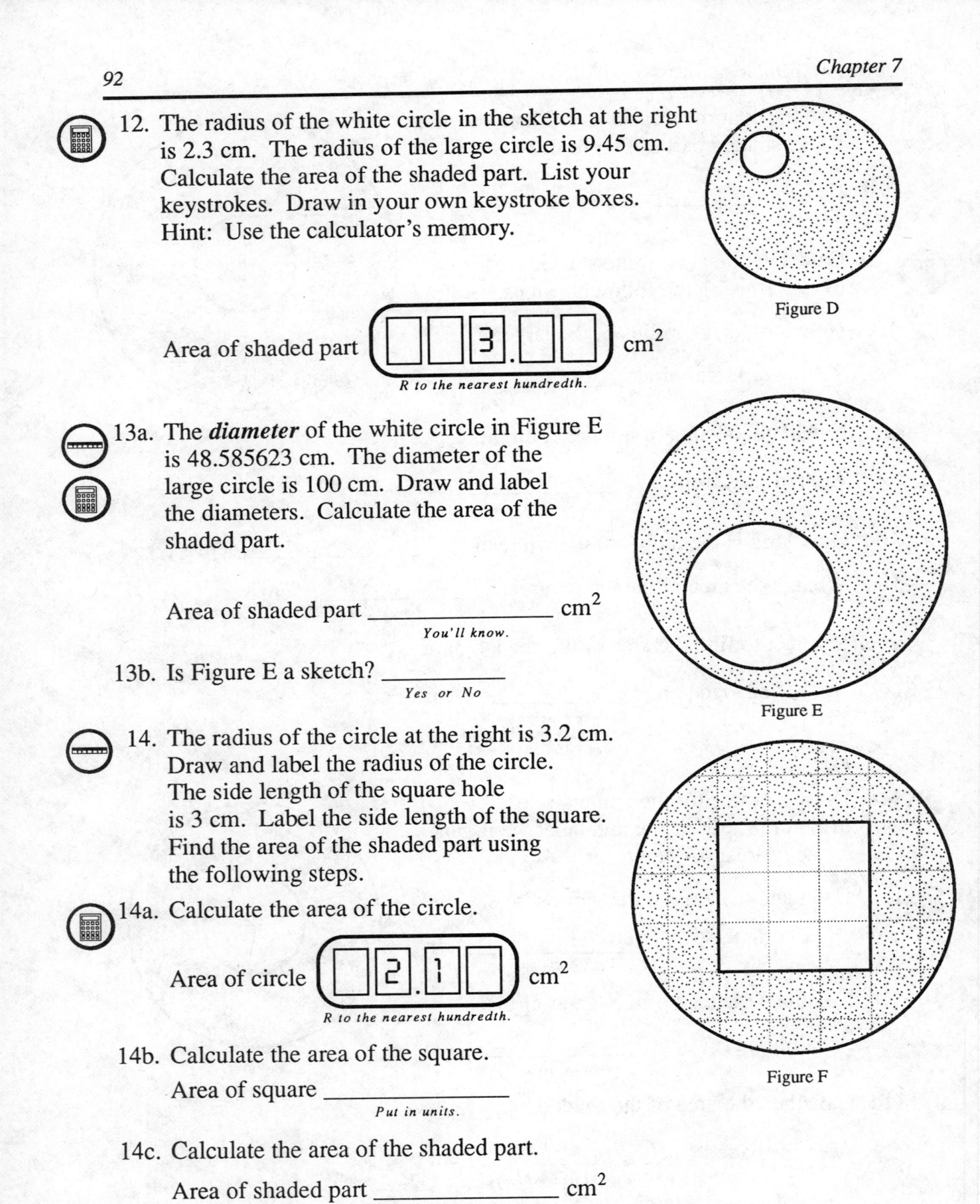

Area of shaded part ☐☐3.☐☐ cm²
R to the nearest hundredth.

Figure D

13a. The *diameter* of the white circle in Figure E is 48.585623 cm. The diameter of the large circle is 100 cm. Draw and label the diameters. Calculate the area of the shaded part.

Area of shaded part _____ cm²
You'll know.

13b. Is Figure E a sketch? _____
Yes or No

Figure E

14. The radius of the circle at the right is 3.2 cm. Draw and label the radius of the circle. The side length of the square hole is 3 cm. Label the side length of the square. Find the area of the shaded part using the following steps.

14a. Calculate the area of the circle.

Area of circle ☐2.1☐ cm²
R to the nearest hundredth.

14b. Calculate the area of the square.

Area of square _____
Put in units.

Figure F

14c. Calculate the area of the shaded part.

Area of shaded part _____ cm²
R to the nearest tenth.

Maneuvers with Circles © David A. Page

Circles with Holes 93

15. The sketch at the right shows a circle with a square hole. The diameter of the circle is 110 cm. Calculate the area of the shaded part. Show your work.

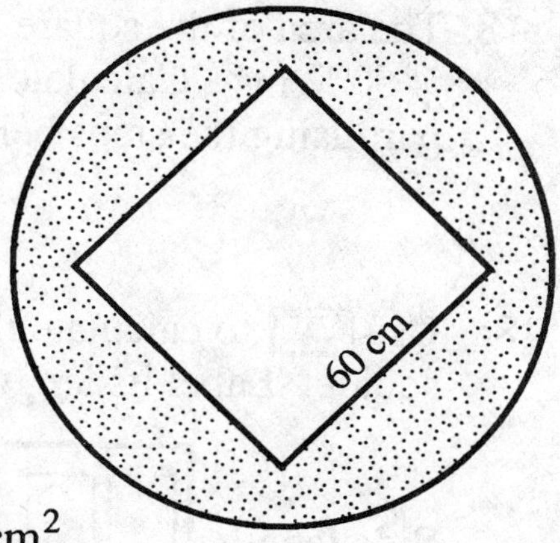

    Area of shaded part ☐ 9 ☐ 3 . ☐ cm²
    R to the nearest tenth.

    Figure G

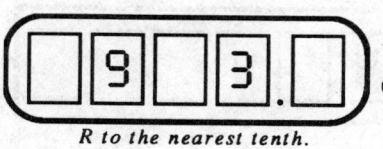

16. The hole in Figure H is a square. Make the necessary measurements to calculate the area of the shaded part. Measure to the nearest 0.1 cm. Label your measurements on the figure and show your work.

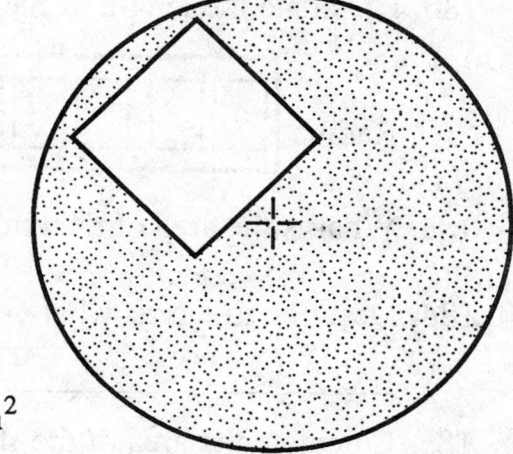

    Area of shaded part _____ cm²
    R to the nearest tenth.

    Figure H

17. There are two square holes in the figure at the right. Calculate the area of the shaded part. Show your work.

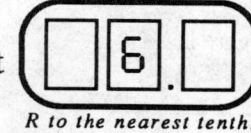

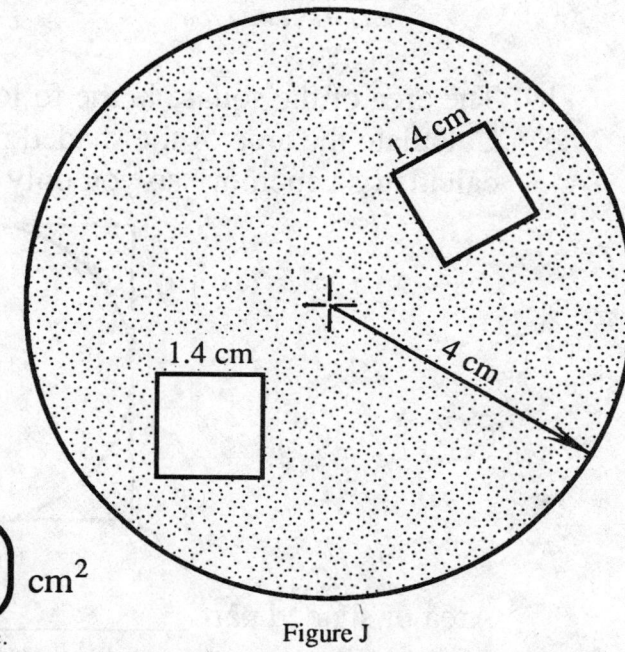

    Area of shaded part ☐ 6 . ☐ cm²
    R to the nearest tenth.

    Figure J

© David A. Page

Maneuvers with Circles

18. The area of the square in the sketch at the right is 600 cm². Calculate the area of the shaded part using the steps below.

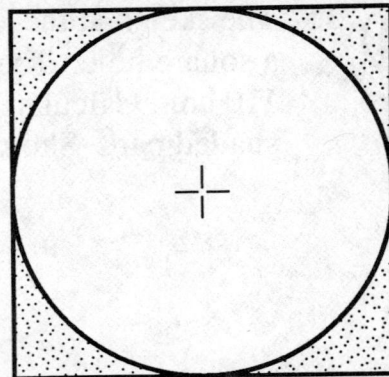

18a. Use $\boxed{\sqrt{x}}$ to calculate the side length of the square. Label its length on the figure.

Side length ☐☐.4☐4☐☐ cm

Figure K

18b. Calculate the length of the radius. Label it on the figure.

Radius ☐☐.☐4☐4☐☐ cm

18c. What is the area of the circle? _____ cm²
Copy window.

18d. Describe in words what you subtract to find the area of the shaded part.

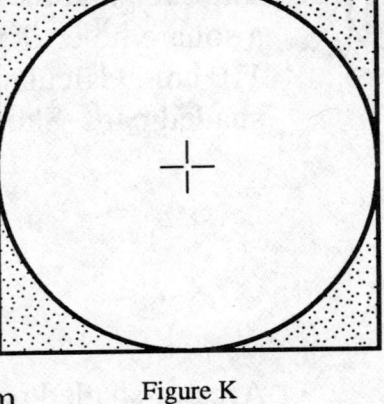

_____ − _____ = Shaded Area

18e. Calculate the area of the shaded part above.

Area _____ cm²
R to the nearest tenth.

19. The area of the square in the following sketch is 573.15446 cm². Calculate the area of the shaded part. Show your work. Use your calculator's memory so you only have to key in the messy number once.

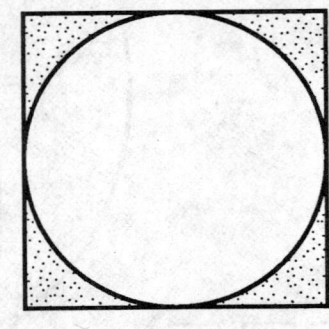

Figure L

Area of shaded part _____ cm²
You'll know.

Maneuvers with Circles © David A. Page

Circles with Holes

20. The area of the square at the right is 10 cm². The area of the shading is 20 cm². Label the figure. What is the area of the circle?

    Area of circle _____
    Put in units.

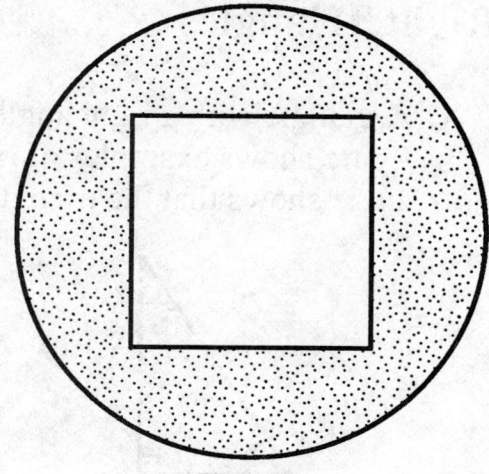

    Figure M

★ 21. The area of the square in the following figure is 9 cm². Calculate the area of the shading. Two of the corners of the square are at the centers of the circles. Show your work.

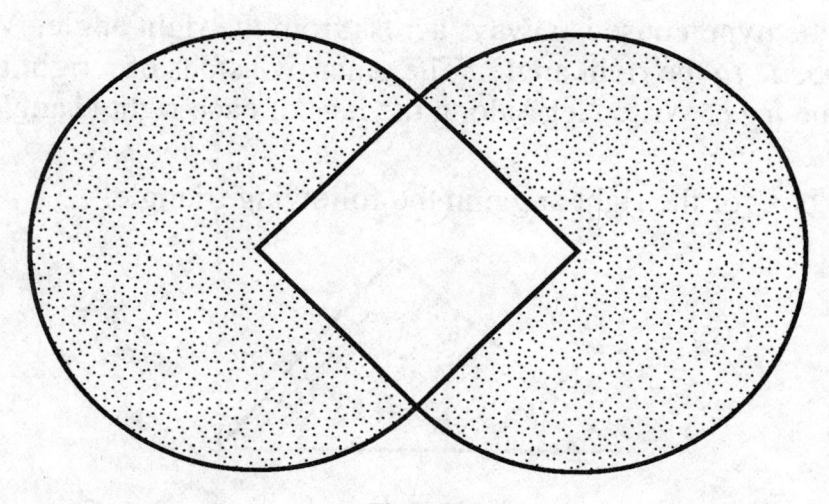

Figure N

Area of shading ☐☐.4☐☐5☐☐ cm²

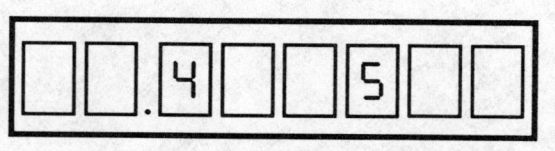

Right Triangles

1. A triangle with a right angle is called a ***right triangle***. The following figure shows examples of right triangles. The "box" drawn in the right angle shows that the triangle is a right triangle.

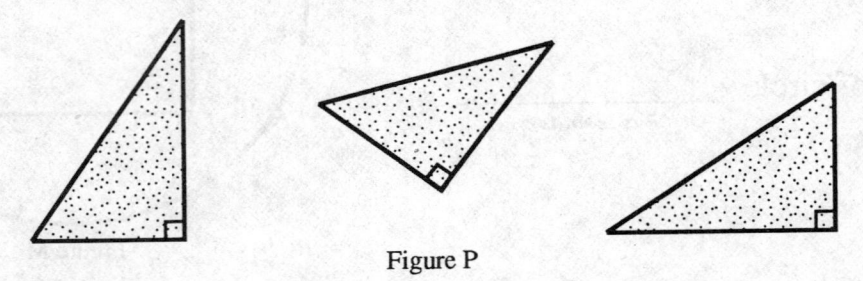

Figure P

1a. The longest side of a right triangle is called the ***hypotenuse***. Write "hypotenuse" along the longest side of each right triangle above.

1b. Notice the hypotenuse is always across from the right angle. We usually say ***opposite to*** the right angle. The other two sides of a right triangle are called the ***legs***. Write "leg" along the legs of each right triangle above.

2a. Draw a box for the right angle in the following triangle.

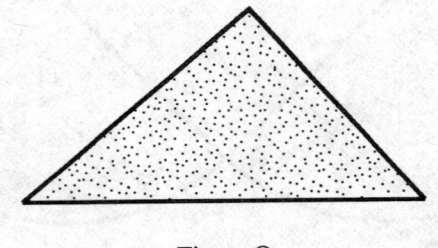

Figure Q

2b. How did you find the right angle? _____

_____

2c. Write "hypotenuse" and "leg" in the correct places in Figure Q.

Maneuvers with Circles © David A. Page

Circles with Holes

A rectangle can be cut into two *congruent* (same size and shape) right triangles.

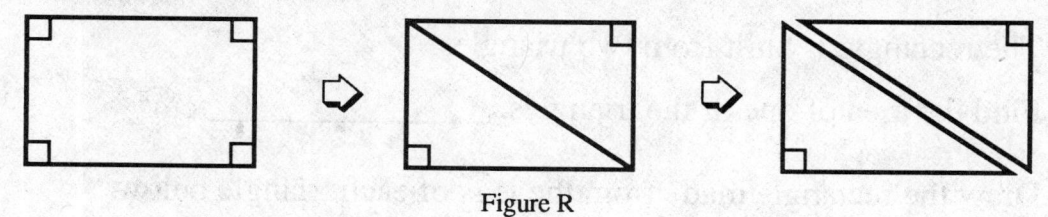

Figure R

Two congruent right triangles build a rectangle.

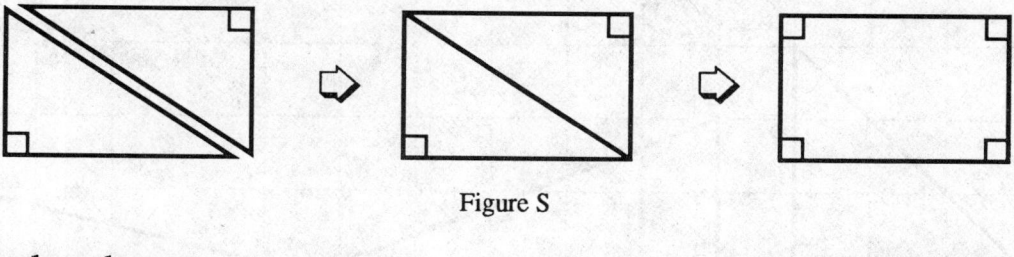

Figure S

To calculate the area of a right triangle, build a rectangle from the two legs as shown in the figure at the right.

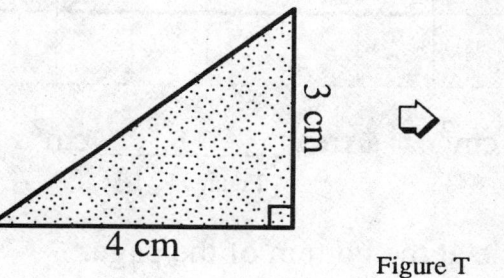

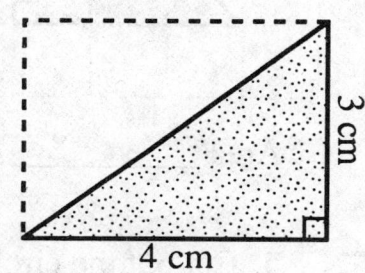

Figure T

3a. What is the area of the rectangle in Figure T? _____ cm²

3b. The rectangle is built from two triangles.

   Find the area of one of the triangles. _____ cm²

4. Draw the rectangle made from the legs of each triangle below. Find the area of each triangle.

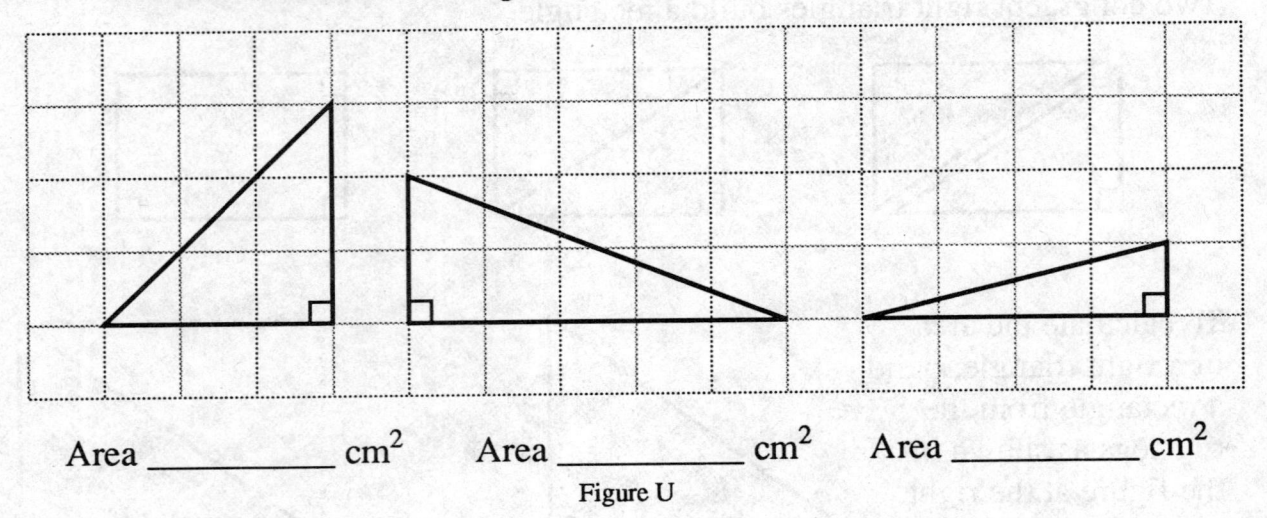

Area _____ cm²   Area _____ cm²   Area _____ cm²

Figure U

5. Cut out the circle at the bottom of the page.

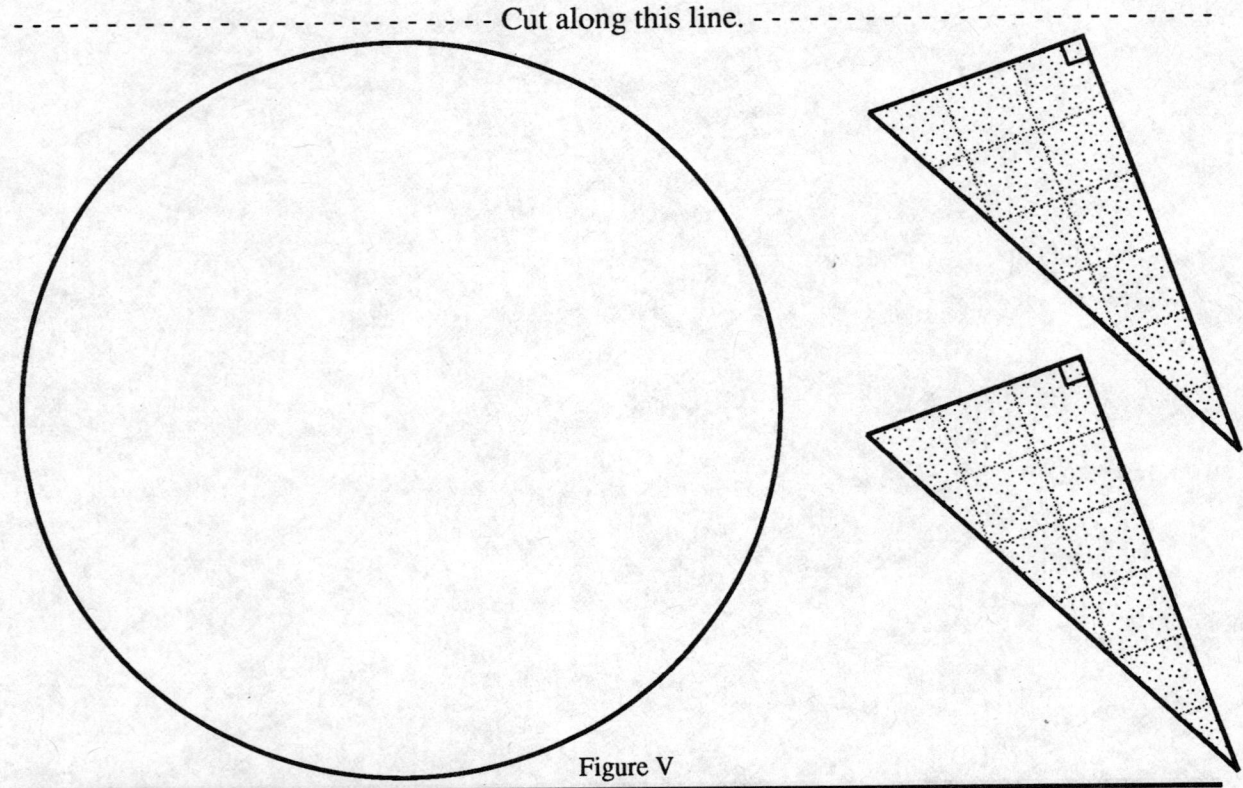

- Cut along this line. -

Figure V

Maneuvers with Circles   © *David A. Page*

Circles with Holes

6. Make the necessary measurements to calculate the area of the circle.

    Area of circle _____ cm²
    <small>R to the nearest tenth.</small>

7. Cut out the right triangles. Compare the triangles with each other.
    What do you notice? _____

8. Place the triangles in the rectangle at the right. The triangles make a rectangle 6 cm by 3 cm. Find the area of the rectangle.

    Area of rectangle _____ cm²

    Figure W

9. The area of the rectangle is equal to the area of 2 triangles.
    What is the area of one of the triangles? _____ cm²

10. Trace one of the triangles inside the circle. Then cut out the triangle.

11. Glue the circle with the triangular hole here.

                                    Glue or tape here.

12. Calculate the area of the part left over.
    Show your work.

    Area _____ cm²
    <small>R to the nearest tenth.</small>

© David A. Page   *Maneuvers with Circles*

13. The diameter of the circle in the sketch at the right is 8 cm.

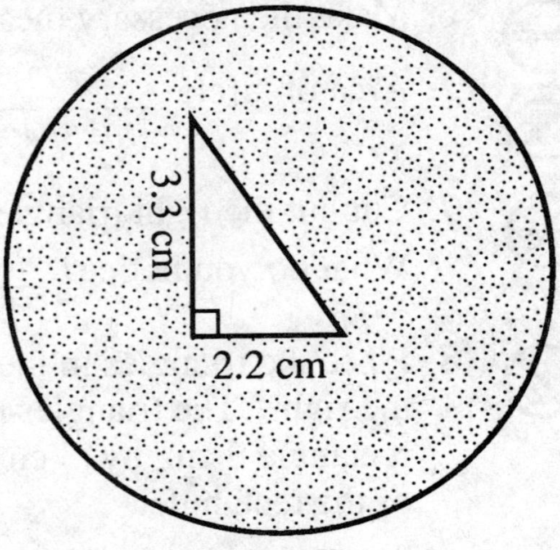

13a. Calculate the area of the circle.

　Area _____ cm²
　　Copy window.

13b. Calculate the area of the triangle.

　Area of triangle _____ cm²

13c. Calculate the area of the shaded part.

　Area of shaded part ☐☐.☐3☐☐☐ cm²

Figure X

14a. Draw a right triangle with legs of 6 cm and 2 cm inside the circle at the right.

14b. Find the area of your triangle and write your answer inside the triangle.

14c. Shade the area outside the triangle.

14d. The shaded area is 38 cm². What is the area of the circle?

　Area of circle _____
　　Put in units.

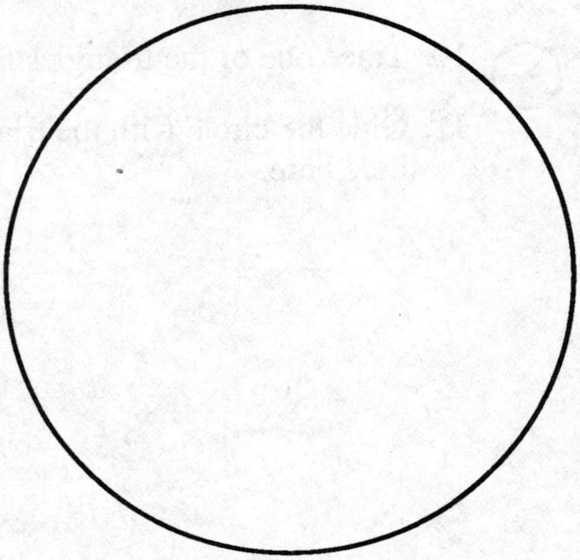

Figure Y

15. Calculate the area of the white part in the sketch at the right. The diameter of the circle is 9.2 cm. Show your work.

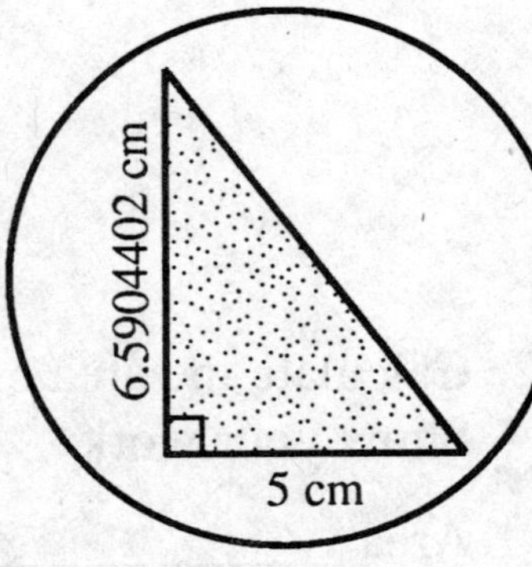

　Area of white part _____ cm²
　　You'll know.

Figure Z

Maneuvers with Circles © *David A. Page*

Circles with Holes 101

16. Estimate the area of the shaded part in Figure AA at the right.

   Estimate _____
   Put in units.

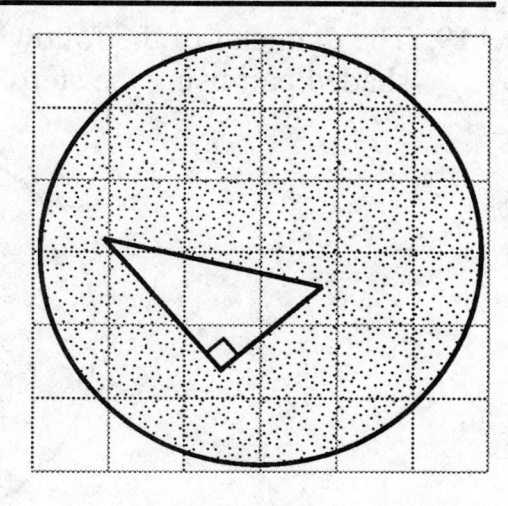

17a. The diameter of the circle at the right is 5.81 cm. Calculate the area of the circle.

   Area of circle (☐ ☐ . 5 ☐) cm²
   R to the nearest hundredth.

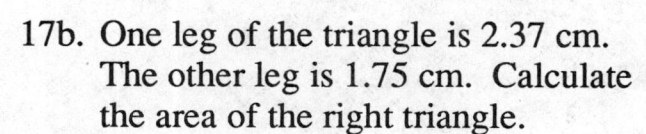

Figure AA

17b. One leg of the triangle is 2.37 cm. The other leg is 1.75 cm. Calculate the area of the right triangle.

   Area of triangle _____ cm²
   R to the nearest hundredth.

17c. Calculate the area of the shaded part.

   Area of shaded part _____ cm²
   R to the nearest tenth.
   Compare with your estimate.

18. Find the area of the circle in Figure BB using the following steps.

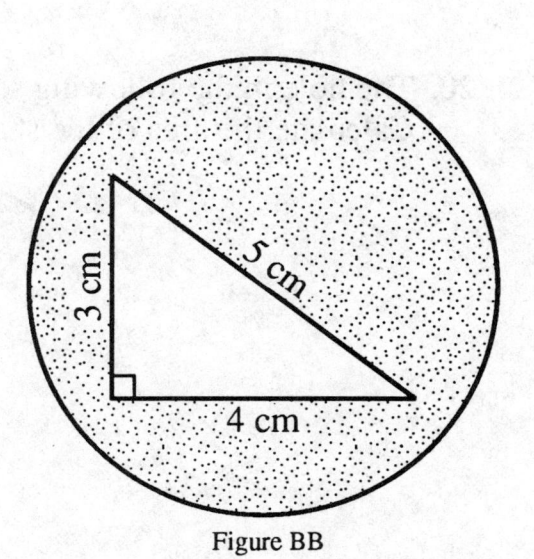

18a. Build a rectangle from the two legs.

18b. Which numbers should you use to find the area of the triangle?

   Answers _____ and _____

18c. What is the area of the triangle?

   Area _____
   Put in units.

18d. The area of the shaded part is 24 cm². What is the area of the circle?

   Area _____
   Put in units.

Figure BB

© David A. Page *Maneuvers with Circles*

19. The diameter of the following circle is 5.5 cm. Calculate the area of the shaded part using the steps below.

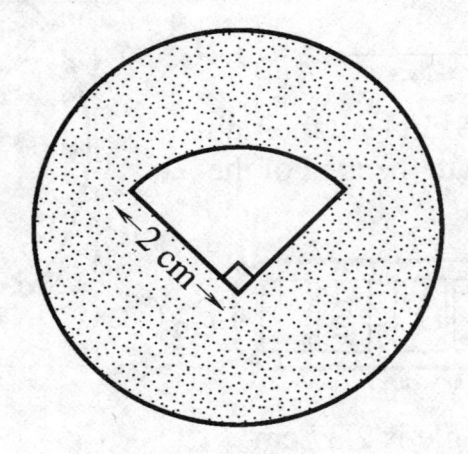

Figure CC

19a. Area of circle _____ cm²
   Copy window.

19b. Area of quarter-circle ☐.☐4☐5☐☐ cm²

19c. Area of shaded part _____ cm²
   R to the nearest tenth.

20. The hole in the following sector is a square with a side length of 1 cm. Calculate the area of the shaded part. Show your work.

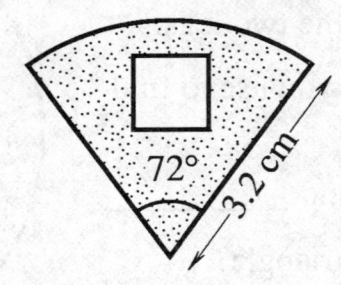

Figure DD

Area of shaded part 5.☐☐3☐☐☐ cm²

Circles with Holes 103

21. Calculate the area of the shaded part in the figure at the right. List your keystrokes using the calculator's memory.

   21a. Area of white circle _____ cm²
   Copy window.

   21b. Area of 60° sector _____ cm²
   Copy window.

   21c. Area of shaded part _____ cm²
   Copy window.

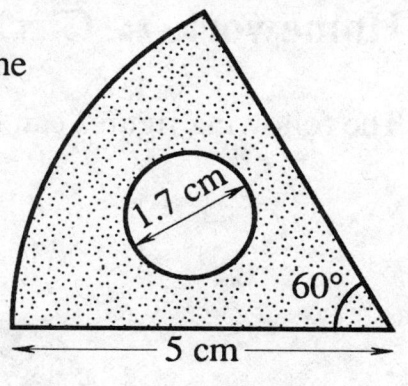

Figure EE

22. Calculate the area of the white part in the figure at the right. Show your work.

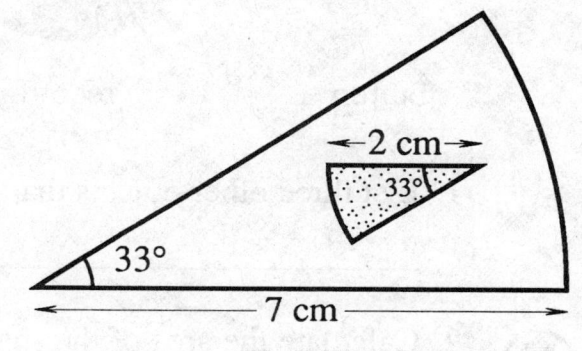

Figure FF

Area of white part ⬜⬜.9⬜ cm²
R to the nearest hundredth.

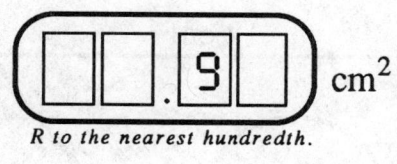

23. The diameter of the circle at the right is 7 cm. Calculate the shaded area. Show your work.

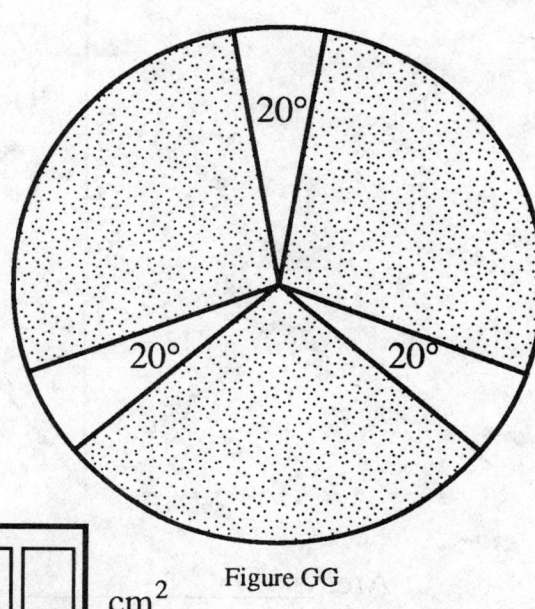

Shaded area ⬜2.⬜⬜⬜⬜2 cm²

Figure GG

© David A. Page   *Maneuvers with Circles*

104 Name _____ Date _____ Class _____

Homework 7: Circles with Holes

The following figures can be thought of as circles with holes in them.

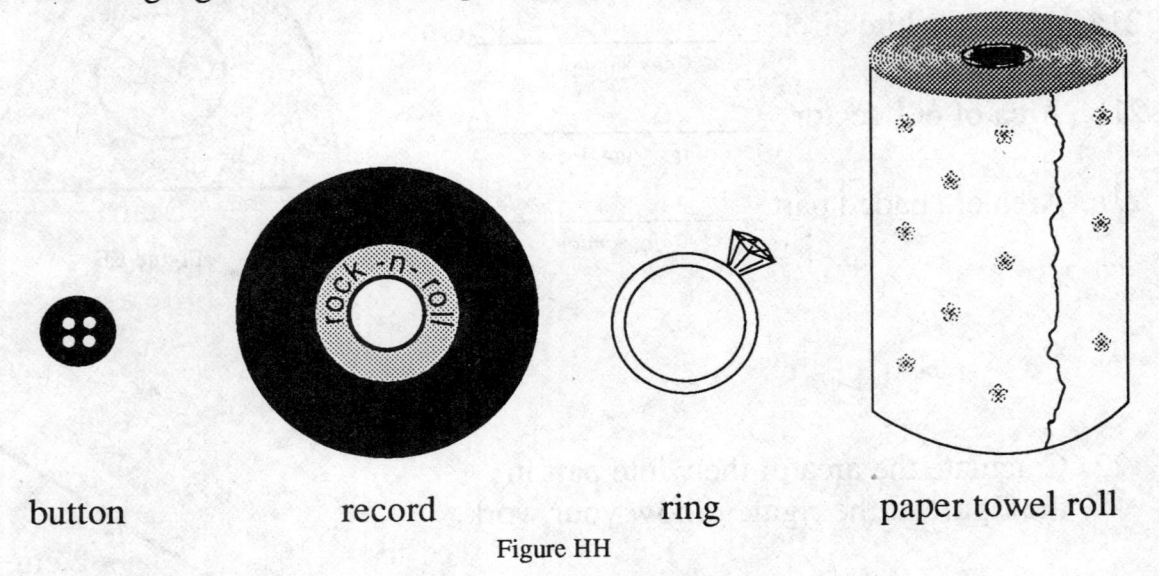

button   record   ring   paper towel roll

Figure HH

1. List three other objects that are circles with holes in them.

   _____ _____ _____

2. Calculate the area of the shaded region in the following sketch.

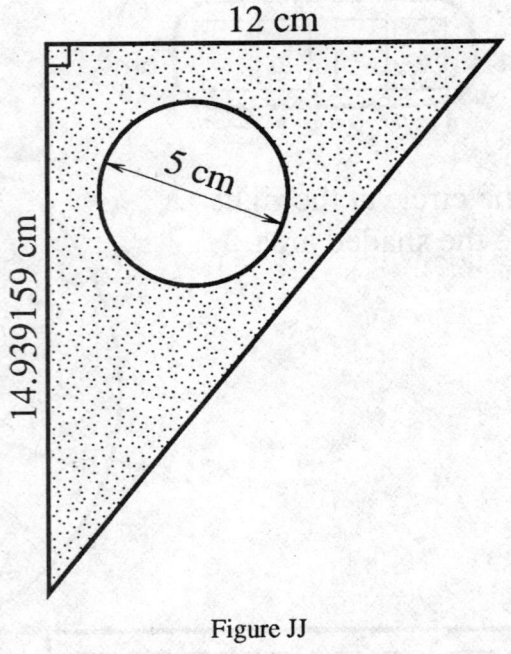

   12 cm

   5 cm

   14.939159 cm

   Figure JJ

   Area _____ cm²
   You'll know.

Maneuvers with Circles   © *David A. Page*

Circles with Holes

3. All the circles in the following figure have a diameter of 2.5 cm. Calculate the area of the shaded part. Use the calculator's memory.

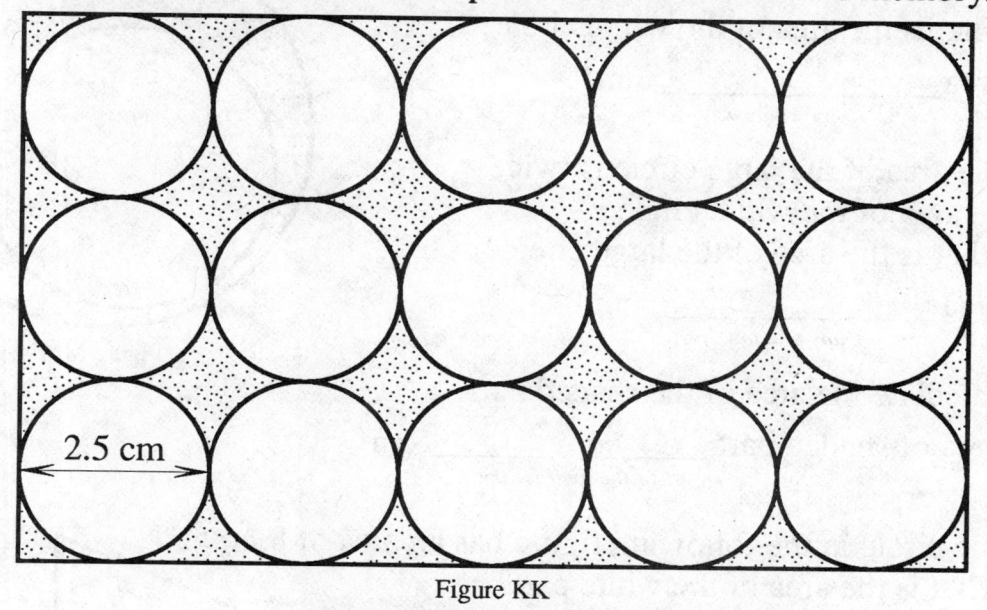

Figure KK

Area of shaded part ☐☐.☐☐☐9☐☐ cm²

4. There are two different-sized circles in the rectangle below. Calculate the area of the shaded part. Use the calculator's memory.

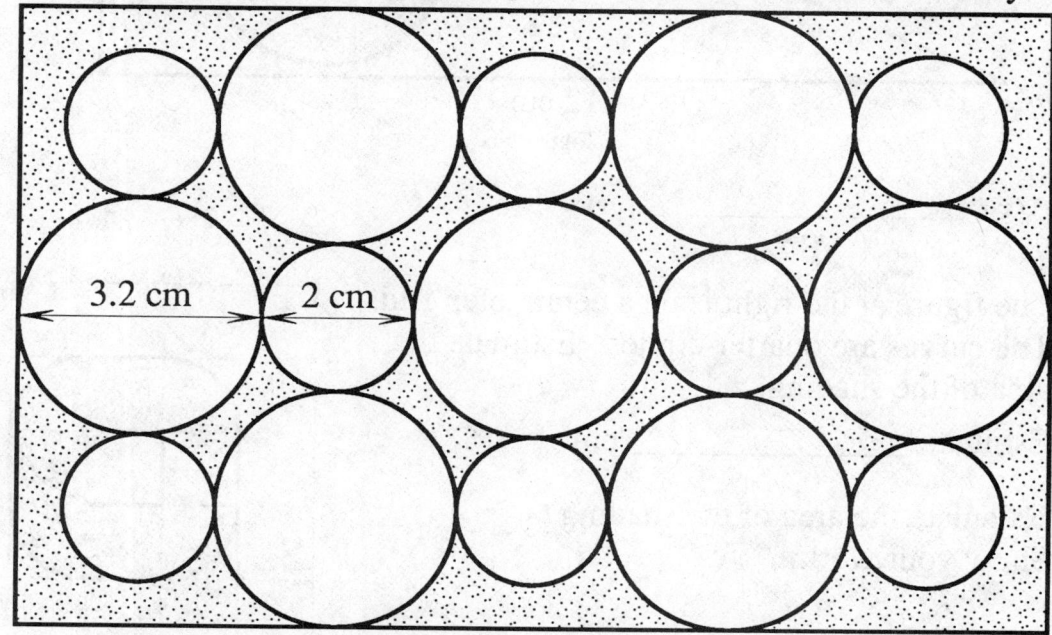

Figure LL

Area ☐☐.8☐9☐☐ cm²

© David A. Page   Maneuvers with Circles

5a. The diameter of the white circle in the sketch at the right is 25.7 cm. What is the area of the white circle?

Area _____ cm²
Copy window.

5b. The area of the large circle is twice the area of the white circle. What is the area of the large circle?

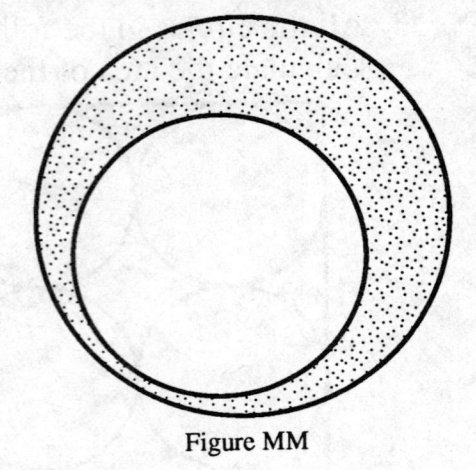

Area _____ cm²
Copy window.

5c. Calculate the area of the shaded part.

Area of shaded part _____ cm²
R to the nearest tenth.

Figure MM

6. The circle in the following figure has an area of 8 cm². What is the area of the white part?

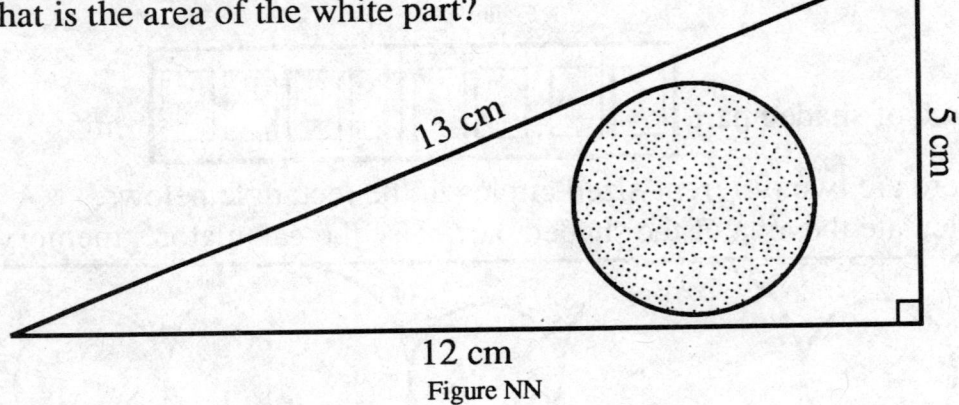

13 cm   5 cm   12 cm

Figure NN

Area _____
Put in units.

7a. The figure at the right is on a centimeter grid. The curves are quarter-circles. Estimate the area of the shading.

Estimate _____ cm²

7b. Calculate the area of the shading. Show your work.

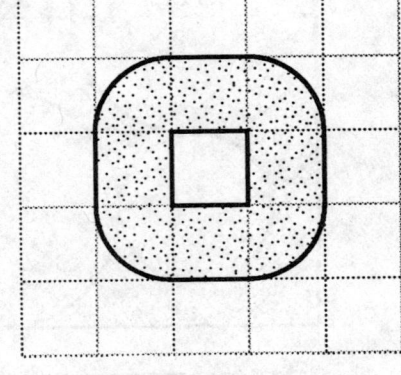

Area | . | 1 | 4 | | | | | cm²

Figure PP

Maneuvers with Circles   © David A. Page

Circles with Holes

107

8. The area of the shading in the figure at the right is 35 cm². The area of the circle is 80 cm². Use the following steps to calculate the side length of the square.

8a. What is the area of the square?

Area of square _____ cm²

8b. Calculate the side length of the square.

Side length of square ☐.☐☐8☐0☐☐

Put in units.

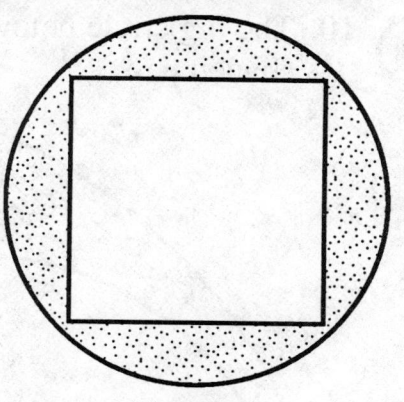

Figure QQ

9. The area of the following square is 49.39804 cm². The semicircle built along its side is cut out of the square. Calculate the area of the shaded region using the steps below.

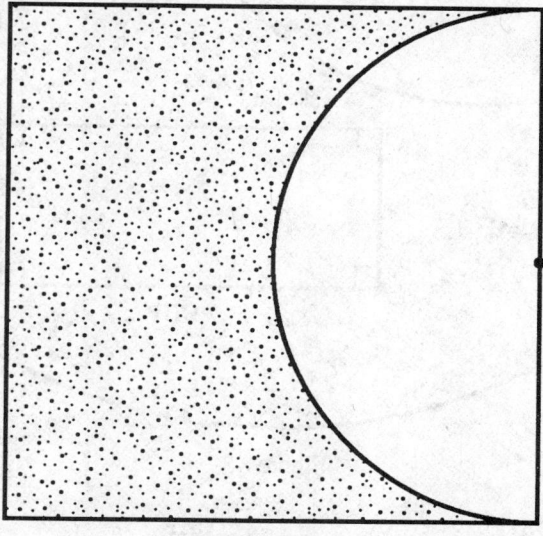

Figure RR

9a. Length of square _____ cm

9b. Radius of semicircle _____ cm

9c. Area of semicircle _____ cm²

9d. Area of shaded part _____ cm²

R to the nearest whole number.

© David A. Page   Maneuvers with Circles

108 Chapter 7

10. The big circle below has a diameter of 15 cm. Calculate the area of the shaded part.

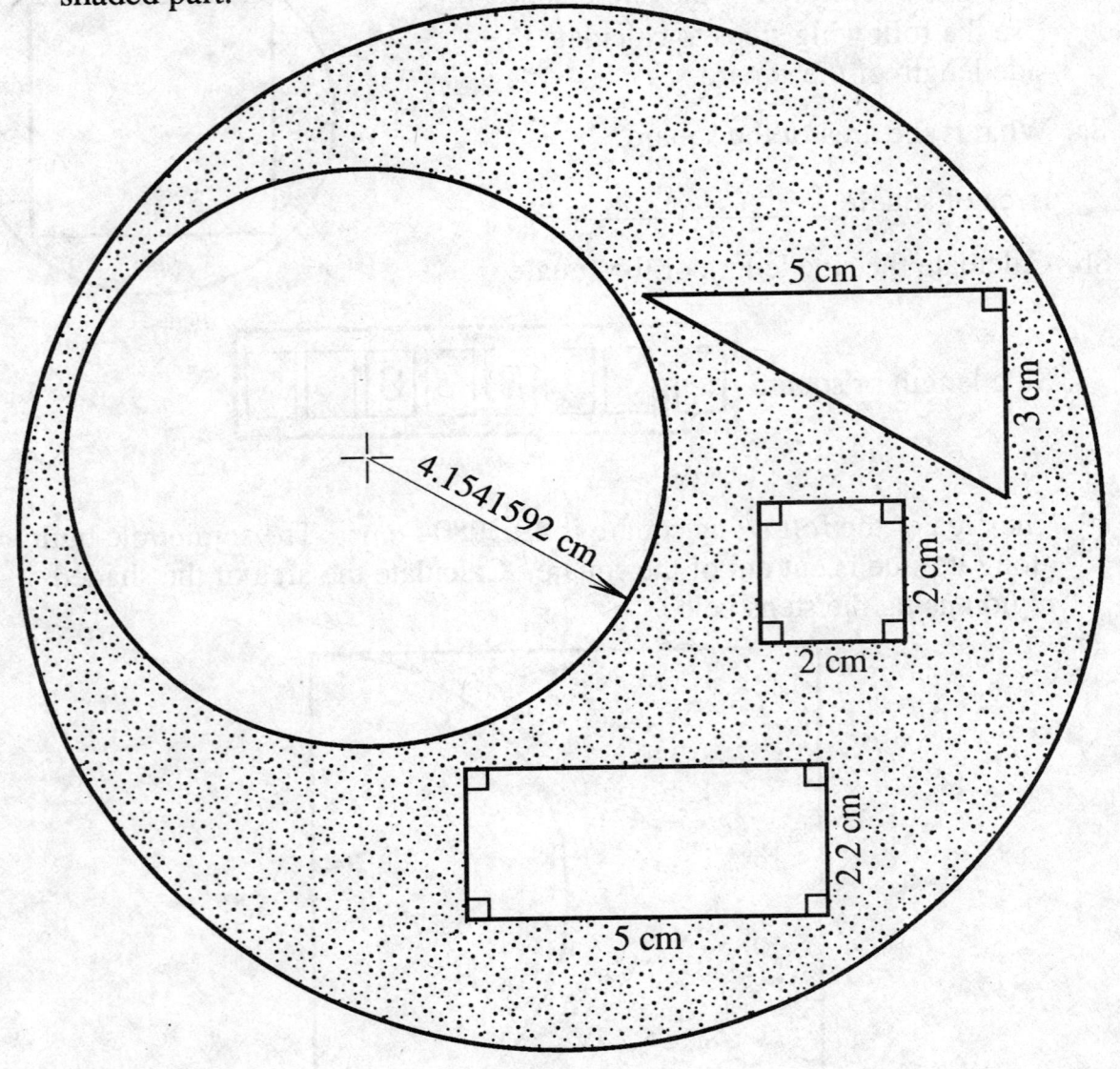

Figure SS

Area of shaded part _____ cm^2
You'll know.

11. The diameter of the semicircle at the right is 6 cm. Calculate the area of the shading.

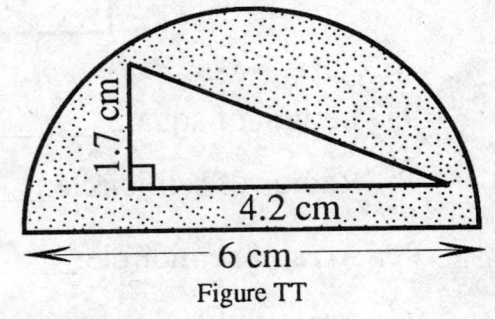

Figure TT

Area of shading [ ][ ].[5][6][ ][ ][ ] cm^2

Maneuvers with Circles © *David A. Page*

Name _____ Date _____ Class _____ 109

8. Perimeter of Pieces

1. The circle at the right has a radius of 3 cm. Use the following steps to calculate the circumference of the circle.

   1a. What is the diameter of the circle?

   Diameter _____ cm

   1b. Calculate the circumference of the circle in Figure A.

   Circumference [☐ ☐ . ☐] cm
   R to the nearest tenth.

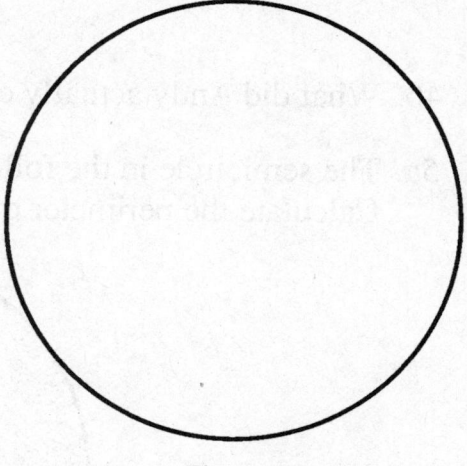

   Figure A

   1c. List your keystrokes.

   | 3 | ☐ | ☐ | ☐ | ☐ | ☐ | ☐ |

   You do not need to use all the keystroke boxes.

2. The semicircle in Figure B also has a radius of 3 cm. Measure the perimeter of the semicircle to the nearest 0.1 cm. Remember, the perimeter is the distance all the way around a shape.

   ←—— Diameter ——→
   Figure B

   Measured perimeter _____ cm

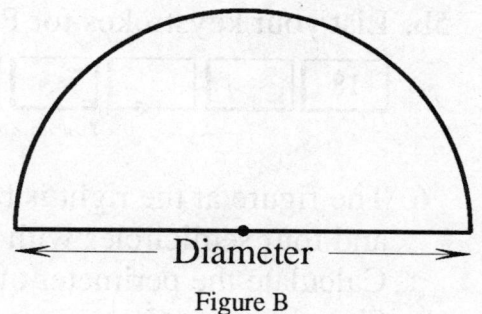

3. Calculate the perimeter of the semicircle in Figure B using the following steps.

   3a. What is the length of the curved part? [☐ . 4 ☐ 4 ☐ ☐] cm

   3b. What is the length of the straight part? _____ cm

   3c. What is the perimeter? _____ cm
   R to the nearest tenth.
   Compare with Problem 2.

© David A. Page   Maneuvers with Circles

110   Chapter 8

4a. To calculate the perimeter of Figure B, Andy used the following keystrokes. Calculate Andy's answer.

| 3 | × | 2 | × | π | ÷ | 2 | = | _____

Copy window.
Compare with Problem 3c.

4b. What did Andy actually calculate? _____

5a. The semicircle in the following sketch has a diameter of 18 cm. Calculate the perimeter of the semicircle.

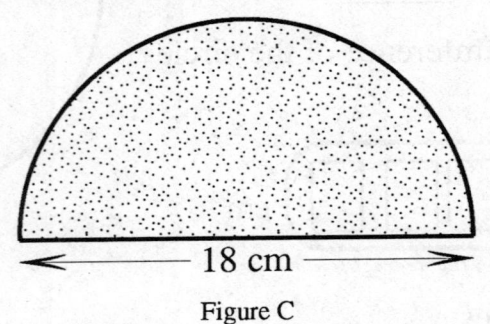

← 18 cm →

Figure C

Perimeter | ☐ | 6 | . | 2 | ☐ | ☐ | ☐ | cm

5b. List your keystrokes for Problem 5a.

| 18 | | | | | | | | | |

You do not need to use all the keystroke boxes.

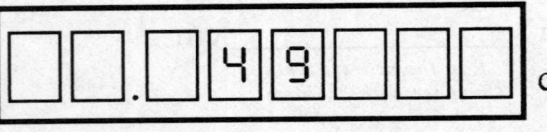

6. The figure at the right is built from a square and four semicircles with diameters of 3 cm. Calculate the perimeter of the figure. Show your work.

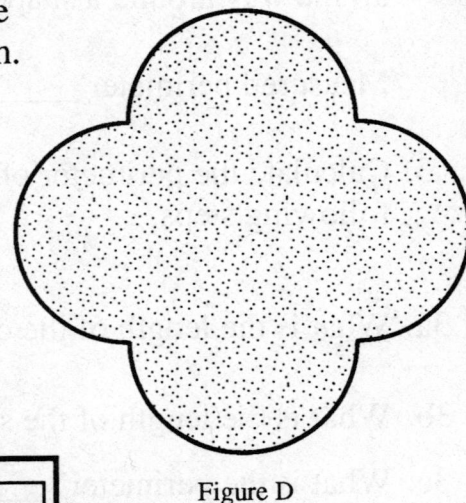

Figure D

Perimeter | ☐ | ☐ | . | 4 | 9 | ☐ | ☐ | cm

Maneuvers with Circles   © *David A. Page*

Perimeter of Pieces 111

7. The sketch at the right is built from four semicircles and a rectangle. The rectangle is 10 cm by 2.7323954 cm.

   Calculate the perimeter (around the outside of the figure) using the following steps and the calculator's memory.

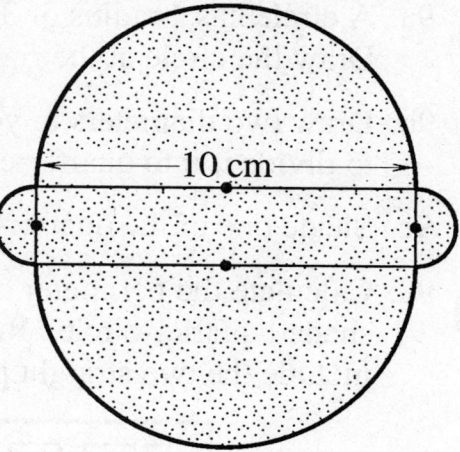

   Figure E

   7a. Calculate the total length of the large curves.

   Length of large curves ☐ 1 . 4 ☐ ☐ ☐ ☐ ☐ cm STO

   7b. Calculate the total length of the small curves.

   Length of small curves _____ cm SUM
   Copy window.

   7c. Press RCL to find the perimeter of the entire figure.

   Answer _____ cm
   You'll know.

8. Calculate the perimeter of the quarter-circle in Figure F using the following steps.

   8a. What is the diameter of the circle?

   Diameter _____ cm

   8b. Calculate the length of the curved part. Remember, four quarter-circles build a circle.

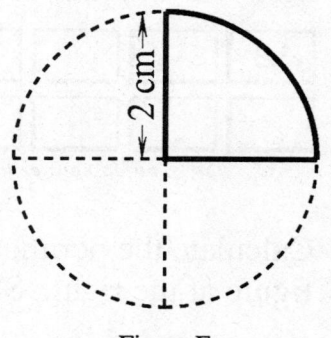

   Figure F

   Length of curved part ☐ . ☐ 4 1 ☐ ☐ ☐ cm

   8c. What is the total length of both straight parts? _____ cm

   8d. What is the perimeter of the quarter-circle? _____ cm
   R to the nearest tenth.

© David A. Page *Maneuvers with Circles*

112

Chapter 8

9a. A circle has a radius of 3.2 cm. Draw the circle at the right.

9b. Draw two diameters in your circle to divide it into quarter-circles.

9c. Shade one of the quarter-circles.

9d. Calculate the perimeter of your shaded quarter-circle. Remember to include the two straight parts.

Perimeter ⬚ 1 . ⬚ 2 ⬚ ⬚ ⬚ cm

10a. The radius of the quarter-circle in Figure G is 3.3 cm. Will the perimeter be smaller or larger than the quarter-circle in Problem 9d?

Answer _____
Smaller or Larger

10b. Calculate the perimeter of this quarter-circle.

Perimeter _____ cm
Compare with Problem 9d.

←— 3.3 cm —→

Figure G

10c. List your keystrokes for Problem 10b.

| 3.3 | | | | | | | |

| | | | | | | | |

You do not need to use all the keystroke boxes.

11. Calculate the perimeter of the shaded figure at the right. Show your work.

Perimeter ⬚ 8 . ⬚ cm
R to the nearest tenth.

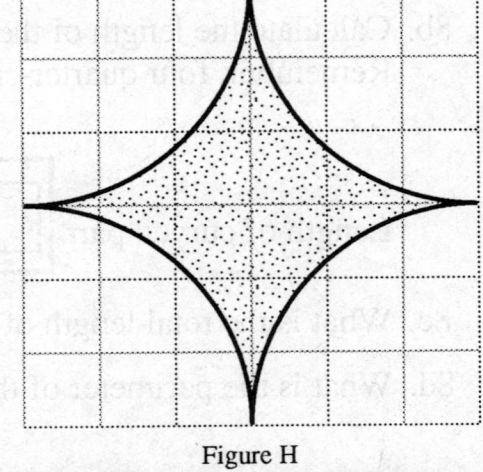

Figure H

Maneuvers with Circles

© David A. Page

Perimeter of Pieces   113

12. The figure at the right is built from four quarter-circles of different sizes. Estimate the perimeter of the figure.

    Estimated perimeter _____ cm

13. Calculate the perimeter of Figure J using the following steps and the calculator's memory. First, calculate the length of each curve. Fill in the missing memory keystrokes.

Figure J

13a. Length of Curve A |_|.|5|_|_|_|_|_| cm   STO

13b. Length of Curve B |_|.|7|_|_|_|_|_| cm |_|

13c. Length of Curve C |_|.|2|_|_|_|_|_| cm |_|

13d. Length of Curve D |_|.|1|_|_|_|_|_| cm |_|

13e. What is the total length of all the straight pieces?

    Total length _____ cm |_|

13f. Press RCL to find the perimeter of the figure.

    Perimeter |_|1|.|_|_|9|_|_| cm

13g. Round your answer to the nearest tenth.

    Perimeter _____ cm

    Compare with Problem 12.

© David A. Page   *Maneuvers with Circles*

114   Chapter 8

14. The diameter of the circle in the sketch at the right is 9 cm. Calculate the circumference of the circle.

    Circumference _____ cm
                   Copy window.

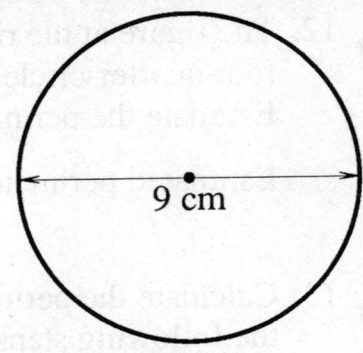

Figure K

15. The sketch at the right is built from a semicircle and three smaller, congruent semicircles. The smaller semicircles lie on the diameter of the large semicircle. Use the following steps to calculate the perimeter of the figure.

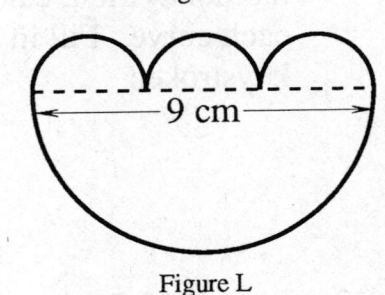

Figure L

15a. Find the length of the large curve.

    Length of large curve ☐☐.1☐7☐☐ cm

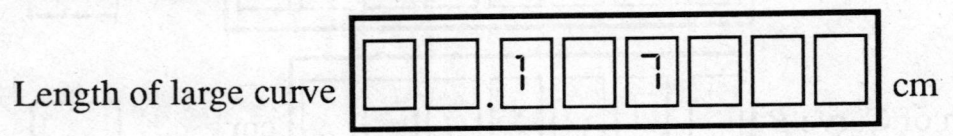

15b. Calculate the total length of the small curves.

    Length of three small curves ☐4.☐3☐☐☐ cm

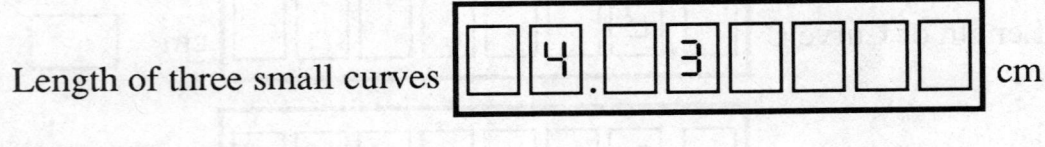

15c. Perimeter of figure _____ cm
                         Copy window.
                    Compare with Problem 14.

16. Three different semicircles lie on the 9 cm diameter in the sketch below. Find the perimeter of the figure using the following steps.

16a. Length of large curve _____ cm
                           Copy window.

16b. Length of small curve _____ cm
                           Copy window.

16c. Length of medium curve _____ cm
                            Copy window.

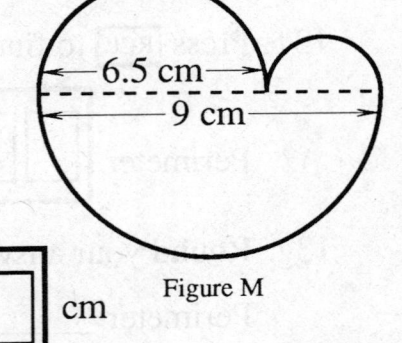

16d. Perimeter of figure ☐8.☐7☐☐☐☐ cm
                    Compare with Problem 15c.

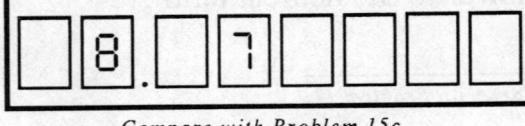

Figure M

Maneuvers with Circles   © *David A. Page*

Perimeter of Pieces 115

Strips

🔲 **1a.** Draw a diameter in the large circle at the bottom of the page.

1b. Measure the diameter to the nearest whole cm. _____ cm

🧭 **2a.** Place the point of your compass on the center of the large circle. Draw a circle with a radius of 3 cm.

2b. What is the diameter of the small circle? _____ cm

3. Shade the part that is outside the small circle.

✂️ **4a.** Cut out the large circle and remove the small circle.

What do you have left? _____

4b. Fold the shaded part in half and cut along the fold.

5. Each of the pieces is a "half-strip". Glue a half-strip in the circle on the next page. A place for the half-strip is drawn in the circle.

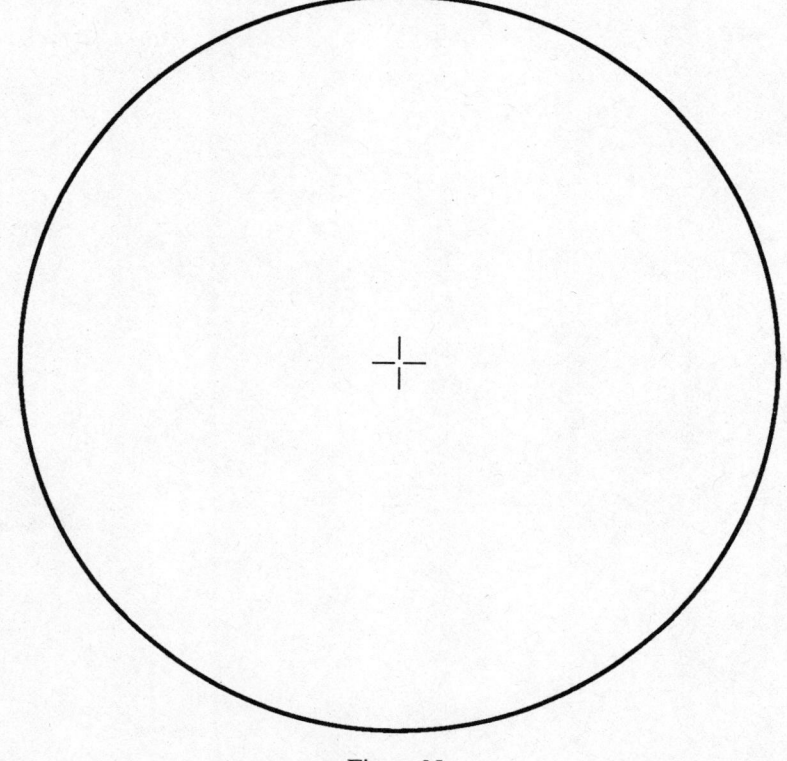

- Cut below this line. -

Figure N

© David A. Page   *Maneuvers with Circles*

116

6. Write the length of the diameter of each circle on the figure at the right.

7. Measure the perimeter of the half-strip to the nearest 0.1 cm.

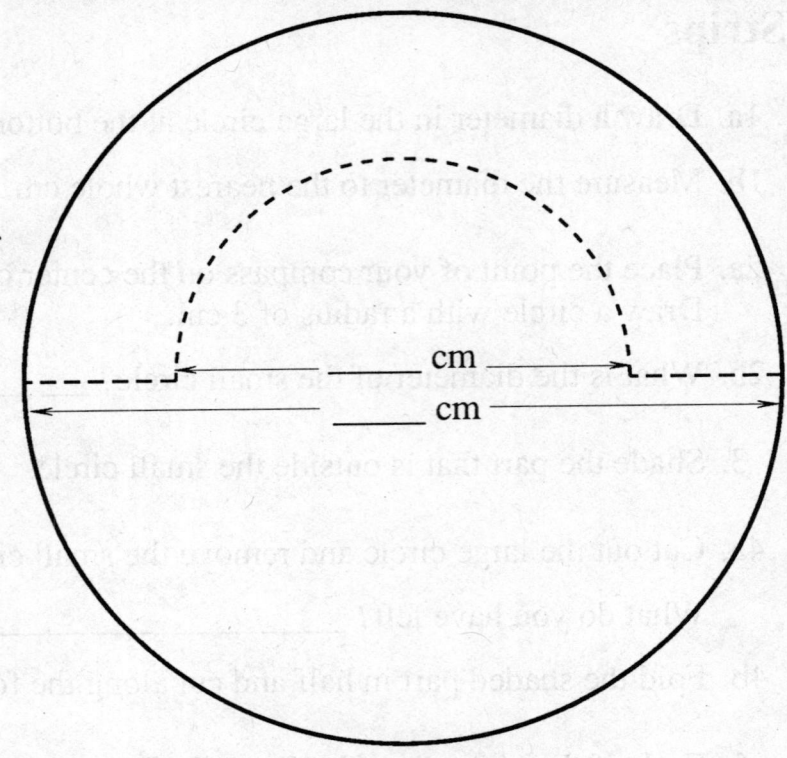

_____ cm
_____ cm

Figure P

Measured perimeter _____ cm

Perimeter of Pieces 117

8. Calculate the perimeter of the half-strip in Figure P using the following steps.

   8a. Length of both straight parts _____ cm

   8b. Calculate the length of the large curve.

   Length of large curve [|5|.| | | | |] cm

   8c. Calculate the length of the small curve.

   Length of small curve [|.|4| |4| | |] cm

   8d. Perimeter [|9|.| |2| | |] cm

   Compare with Problem 7.

9. Use the following steps to calculate the perimeter of the half-strip at the right.

   9a. The large curve is part of a large circle. What is the diameter of the circle?

   Diameter of large circle _____ cm

   9b. Length of large curve [|.|4| |4| | |] cm

   9c. The small curve is part of a smaller circle. What is the diameter of this circle?

   Diameter of small circle _____
   Put in units.

   9d. Length of small curve _____ cm
   Copy window.

   9e. Length of both straight parts _____ cm

   9f. Perimeter [|7|.| |7| | |] cm

Figure Q (1.5 cm, 3 cm)

© David A. Page   *Maneuvers with Circles*

118 Chapter 8

A half-strip is cut in half as shown below. One of these new pieces is called a quarter-strip.

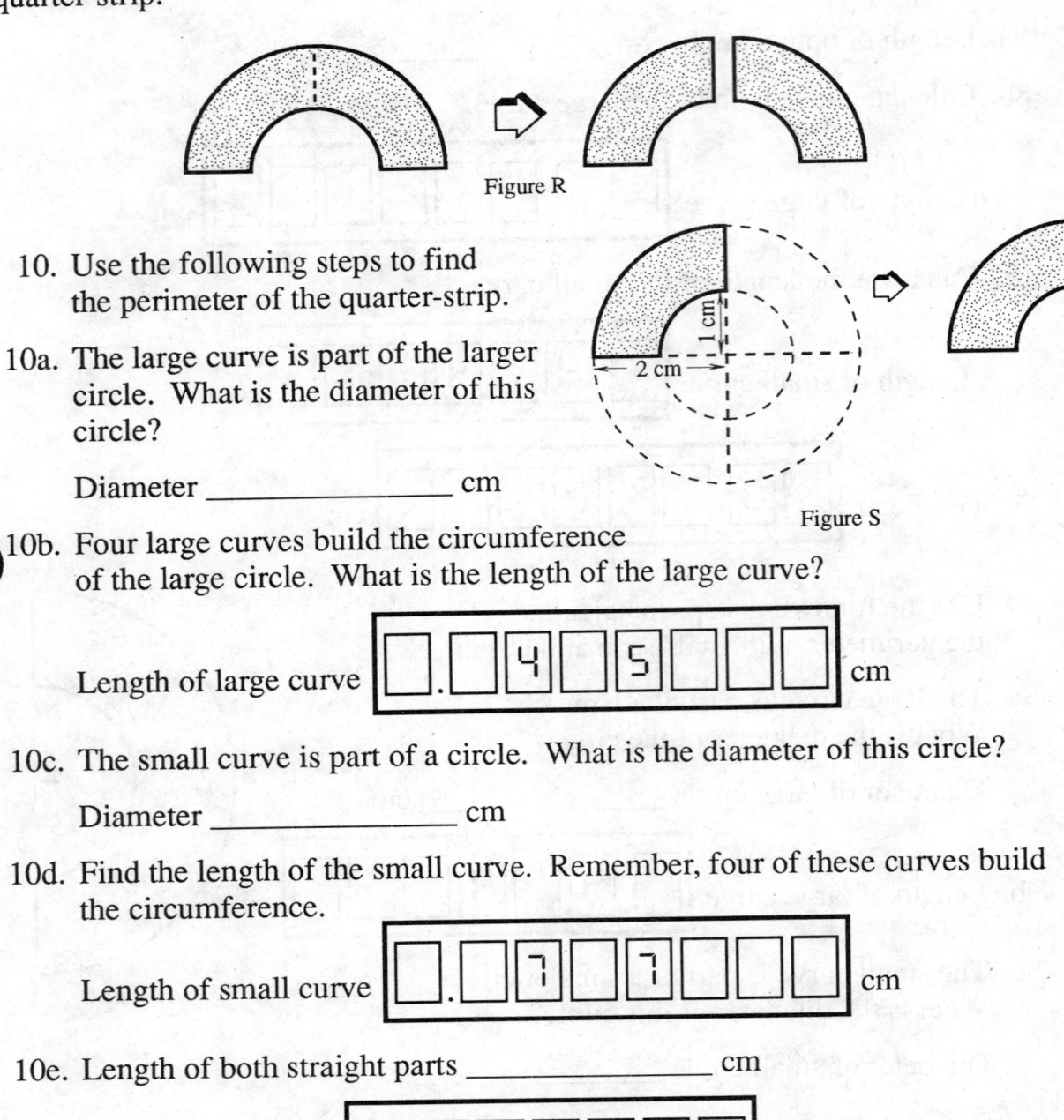

Figure R

Figure S

10. Use the following steps to find the perimeter of the quarter-strip.

10a. The large curve is part of the larger circle. What is the diameter of this circle?

Diameter _____ cm

10b. Four large curves build the circumference of the large circle. What is the length of the large curve?

Length of large curve ☐.☐4☐5☐☐☐ cm

10c. The small curve is part of a circle. What is the diameter of this circle?

Diameter _____ cm

10d. Find the length of the small curve. Remember, four of these curves build the circumference.

Length of small curve ☐.☐7☐7☐☐☐ cm

10e. Length of both straight parts _____ cm

10f. Perimeter of figure ☐.☐7☐2☐☐ cm

Maneuvers with Circles © *David A. Page*

Perimeter of Pieces   *119*

11. Estimate the perimeter of the following quarter-strip.

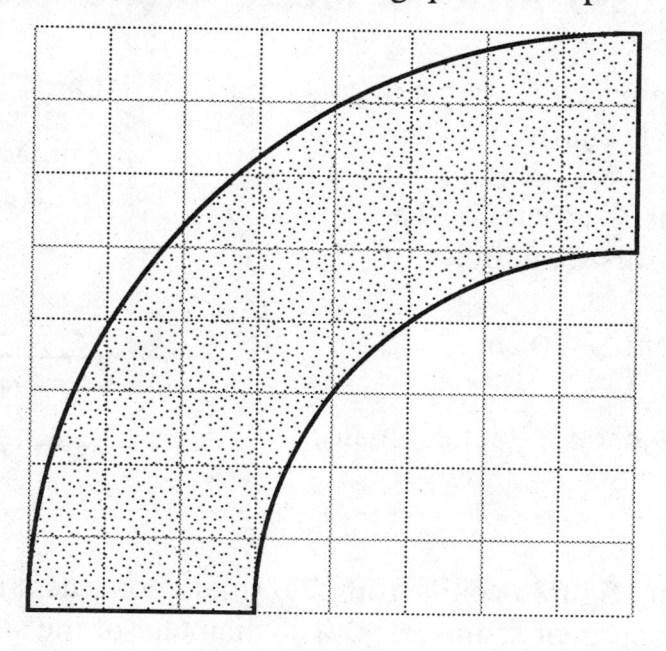

Figure T

   Estimated perimeter _____ cm

12. Calculate the perimeter of the quarter-strip using the following steps.

   12a. Length of large curve ☐☐.☐6☐6☐☐☐ cm

   12b. Length of small curve ☐.8☐☐☐8☐☐ cm

   12c. Length of both straight parts _____ cm

   12d. Perimeter _____ cm
   R to the nearest tenth.
   Compare with your estimate.

© *David A. Page*   *Maneuvers with Circles*

120 Name _____ Date _____ Class _____

Homework 8: Perimeter of Pieces

1a. The circumference of a circle is 24.4 cm. A semicircle is cut from this circle as shown in the sketch at the right. Circle the best estimate for the perimeter of this semicircle.

   5 cm   10 cm   20 cm   25 cm

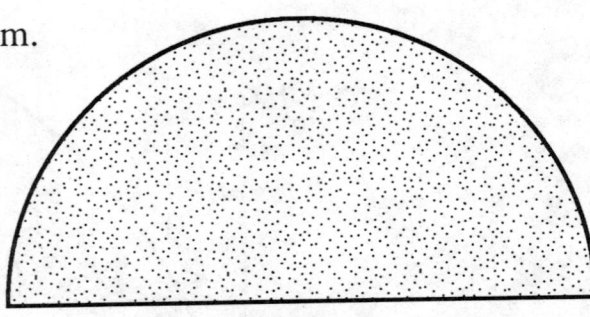

Figure U

1b. How did you choose your estimate? _____

_____

2. The following figure is built from two small, congruent semicircles and two large, congruent semicircles. The diameter of the small semicircle is 3 cm. The diameter of the large semicircle is 5 cm. Calculate the perimeter of the figure using the steps below.

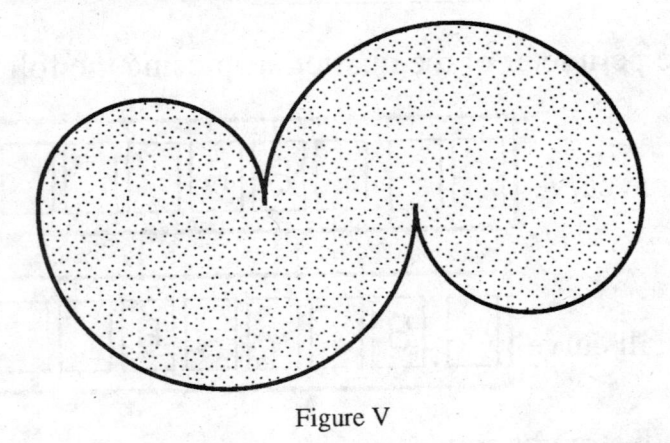

Figure V

2a. Total length of small curves   ☐ . ☐ 2 4 ☐ ☐ cm

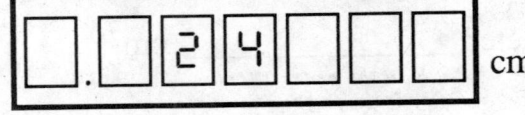

2b. Total length of large curves   ☐ ☐ . 7 ☐ 7 ☐ ☐ cm

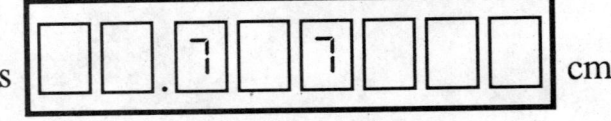

2c. Perimeter   ☐ 5 . ☐ 3 ☐ ☐ ☐ cm

Maneuvers with Circles   © *David A. Page*

Perimeter of Pieces

3. The sketch at the right shows a basketball key. Calculate the perimeter of the basketball key.

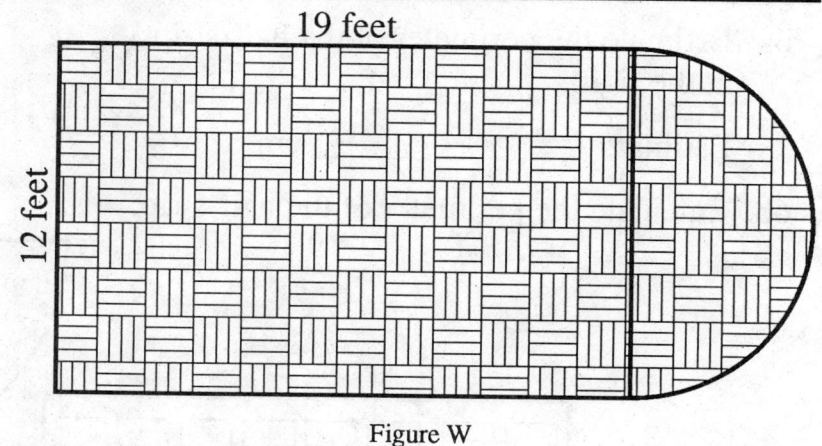

19 feet

12 feet

Figure W

Perimeter ☐ ☐ . 8 ☐ ☐ ☐ ☐ ft

★ 4. The sketch at the right is built from a rectangle and a semicircle. The perimeter of the entire figure is 26.3 cm. Calculate the length of the rectangle.

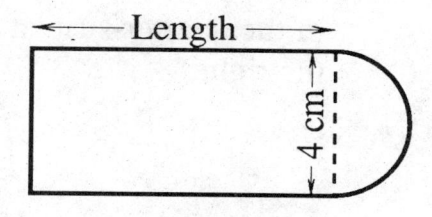

Length

4 cm

Figure X

Length ☐ . ☐ ☐ 8 4 ☐ ☐ cm

5. The figure at the right is built from two quarter-circles. Find its perimeter using the following steps. Label each part as you go along.

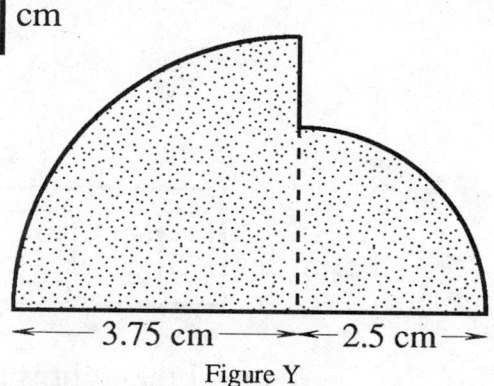

3.75 cm 2.5 cm

Figure Y

5a. Length of large curve ☐ . ☐ ☐ 0 4 ☐ ☐ cm

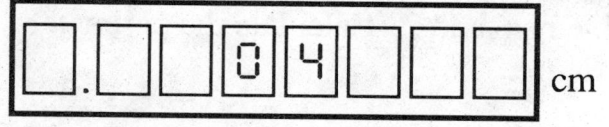

5b. Length of small curve _____ cm

Copy window.

5c. Length of straight parts _____ cm

5d. Perimeter ☐ 7 . 3 ☐ ☐ ☐ ☐ cm

© David A. Page *Maneuvers with Circles*

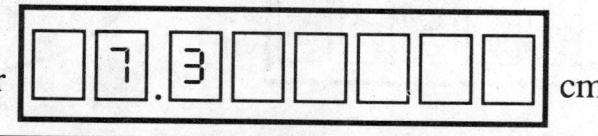

Chapter 8

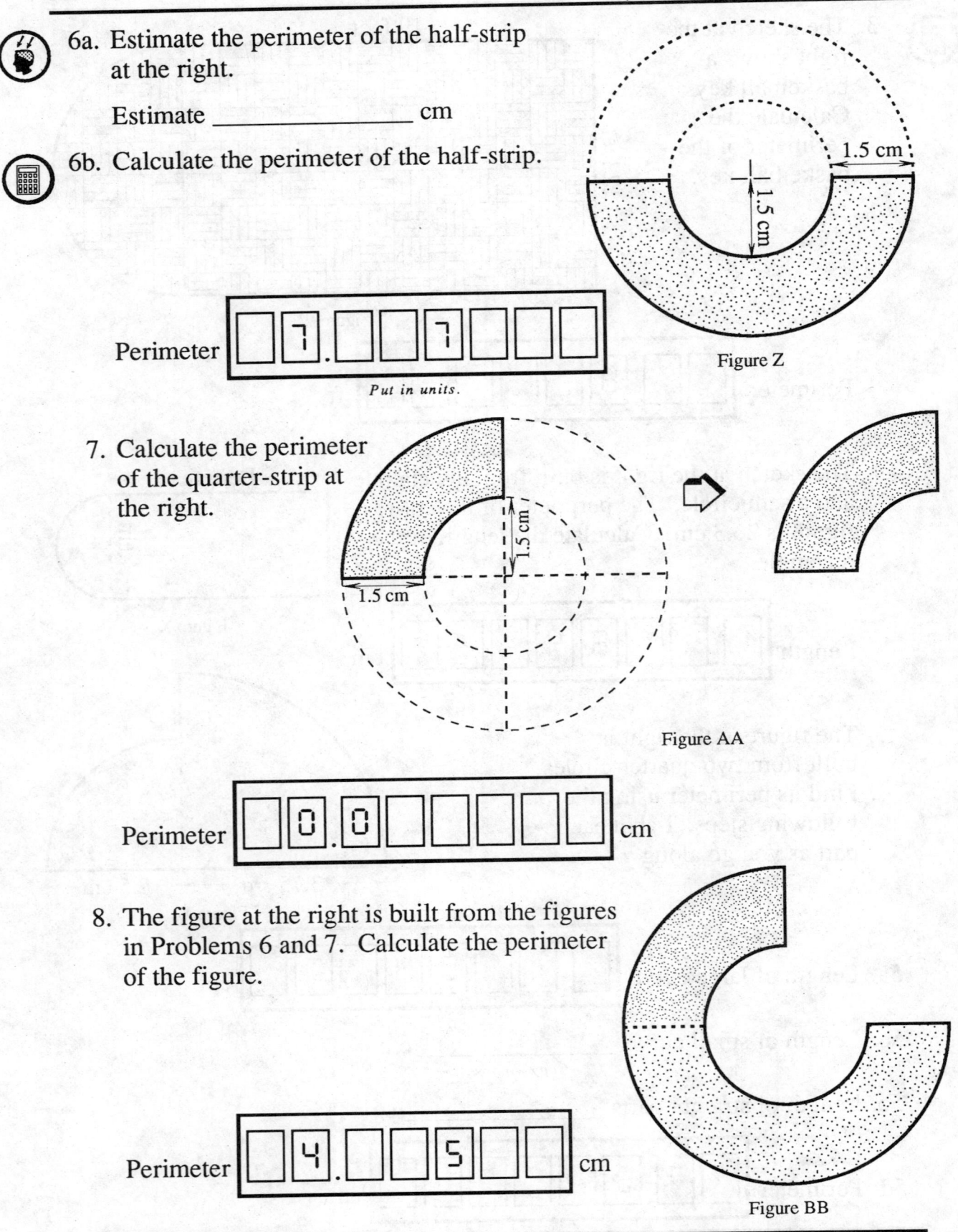

6a. Estimate the perimeter of the half-strip at the right.

Estimate _____ cm

6b. Calculate the perimeter of the half-strip.

Perimeter | 7 | . | | 7 | | | |
Put in units.

Figure Z

7. Calculate the perimeter of the quarter-strip at the right.

Figure AA

Perimeter | | 0 | . | 0 | | | | | cm

8. The figure at the right is built from the figures in Problems 6 and 7. Calculate the perimeter of the figure.

Perimeter | | 4 | . | | 5 | | | cm

Figure BB

Maneuvers with Circles © David A. Page

Name _____ Date _____ Class _____ 123

9. Circles, Chords, and the Pythagorean Theorem

The lines in the circles in Figure A are *chords*.

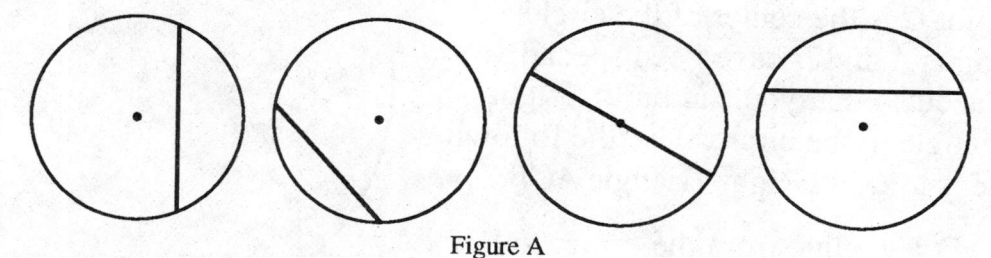

Figure A

The lines in the circles in Figure B are *not* chords.

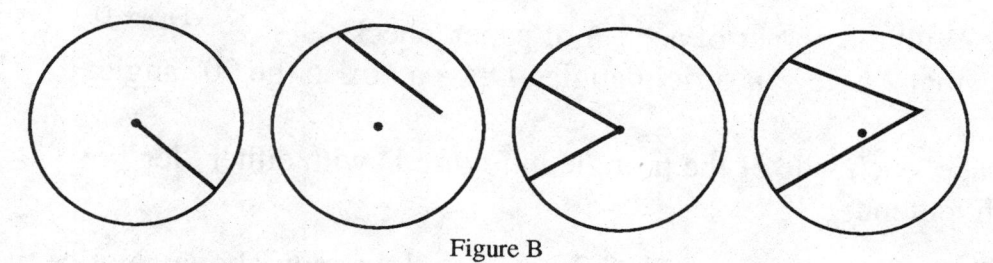

Figure B

1. Describe a chord. _____
   _____

2. Place a check below each circle that has one or more chords drawn in it.

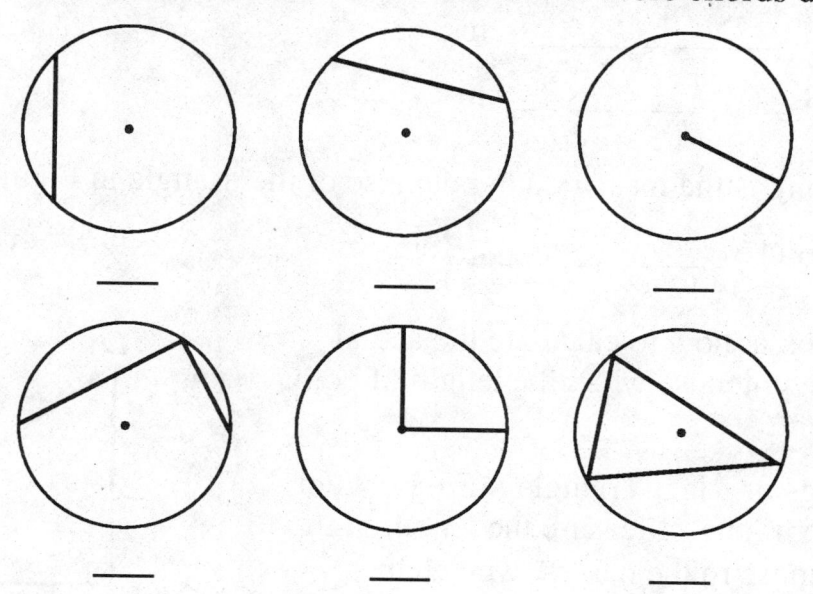

   ___   ___   ___

   ___   ___   ___

   Figure C

© David A. Page

Maneuvers with Circles

3. Measure the length of the chord at the right to the nearest 0.1 cm.

   Measured length _____ cm

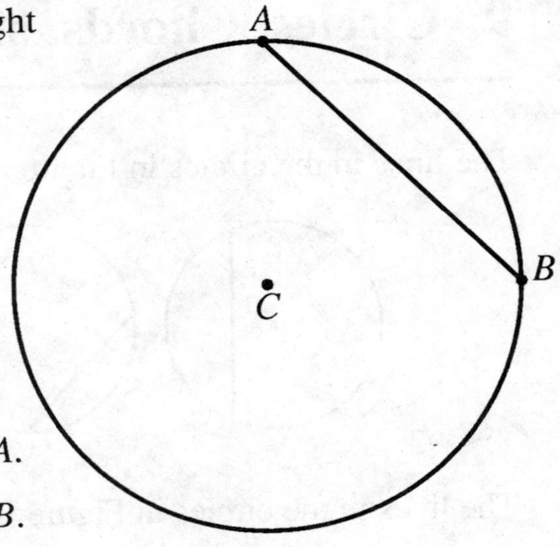

4. Point C is the center of the circle. Points A and B have been specially placed so that you can build a right triangle in the circle. Use the following steps to build Right Triangle ACB.

   a. Draw a line from the center to Point A.

   b. Draw a line from the center to Point B.

   c. With the corner of a sheet of paper, check that Angle C is a right angle. Draw a box in the 90° angle.

Figure D

5a. Label each side of the triangle in Figure D with either "leg" or "hypotenuse."

5b. The radius of the circle is 3.3 cm. Label the radii above. Notice the legs of the triangle are also radii of the circle, and the hypotenuse of the triangle is a chord of the circle.

6. How long is each leg of the triangle?

6a. Leg AC _____ cm

6b. Leg CB _____ cm

7. How long is the measured hypotenuse of the triangle in Figure D?

   Hypotenuse _____ cm
   Should agree with Problem 3.

Now let's look at how to calculate the actual length of a hypotenuse when the length of both legs are given.

8. The legs of Right Triangle DEF are 3 cm and 4 cm long. Measure the length of the hypotenuse to the nearest whole cm.

   Measured hypotenuse _____ cm

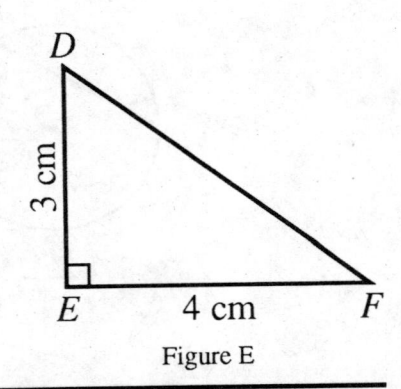

Figure E

Maneuvers with Circles

Circles, Chords, and the Pythagorean Theorem

Pythagoras, a famous Greek mathematician and philosopher, knew a way to calculate the hypotenuse of any right triangle if he knew the length of the two legs.

Pythagoras found the length of the hypotenuse by building squares along each of the sides of the triangle as shown at the right.

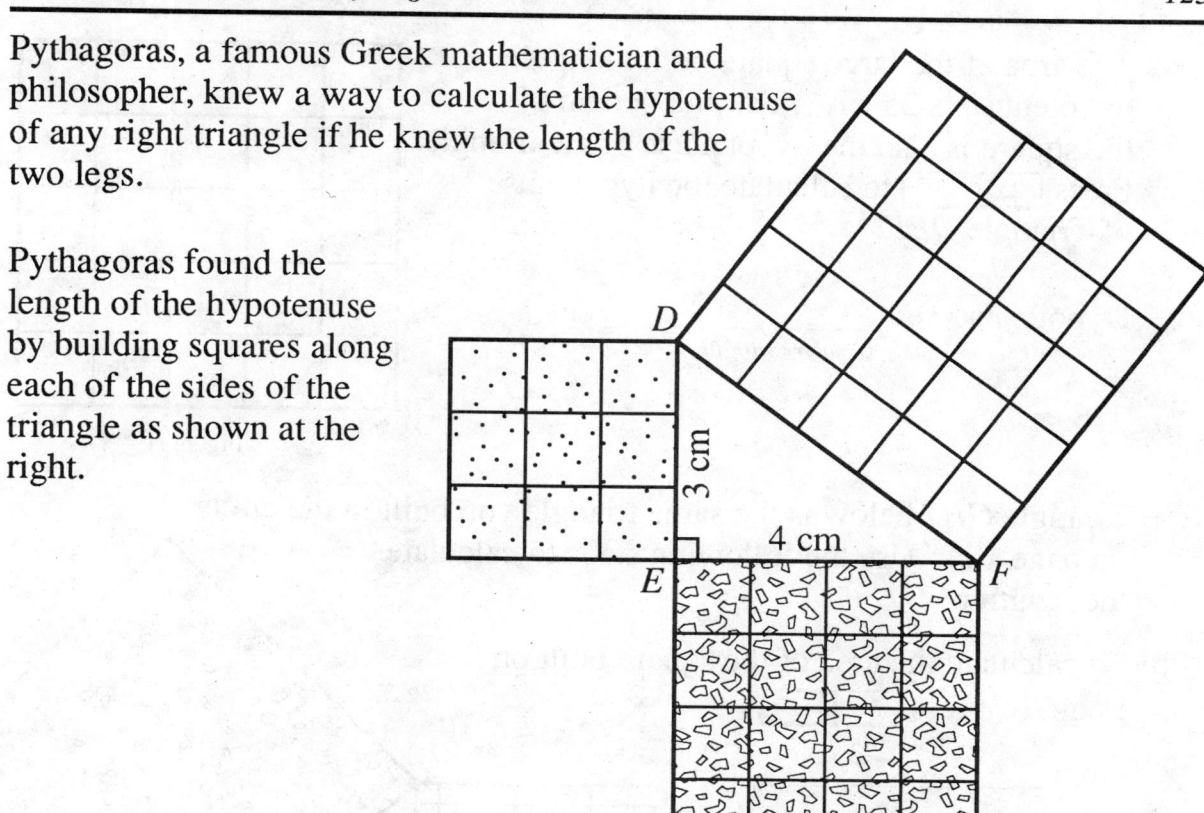

Figure F

Pythagoras noticed that the area of the square built along one leg plus the area of the square built along the other leg equaled the area of the square built along the hypotenuse. This is called the **Pythagorean Theorem**.

| Number of cm² in square built along one leg | + | Number of cm² in square built along the other leg | = | Number of cm² in square built along hypotenuse |

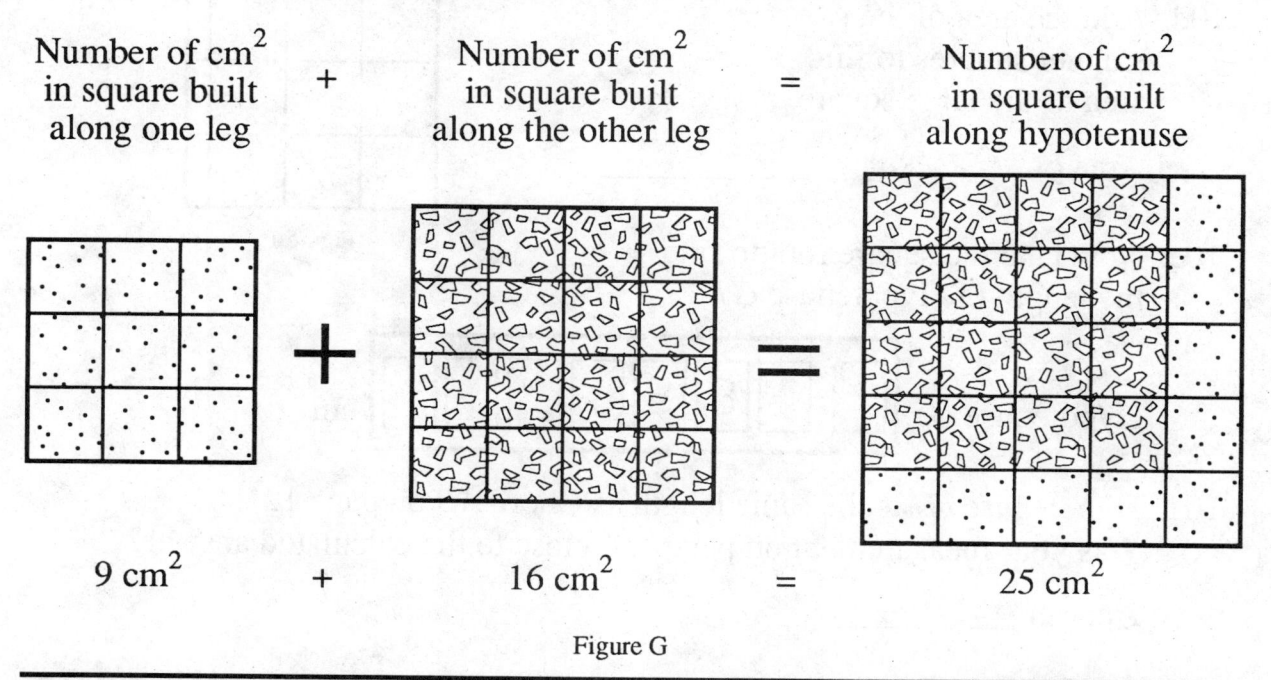

$$9 \text{ cm}^2 + 16 \text{ cm}^2 = 25 \text{ cm}^2$$

Figure G

126   Chapter 9

9. The area of the large square built along the hypotenuse is 25 cm². The side length of the square is also the hypotenuse of the triangle. Press [25] [√x̄] to calculate the hypotenuse of Triangle *DEF*.

   Hypotenuse *DF* _____ cm
   Compare with Problem 8.

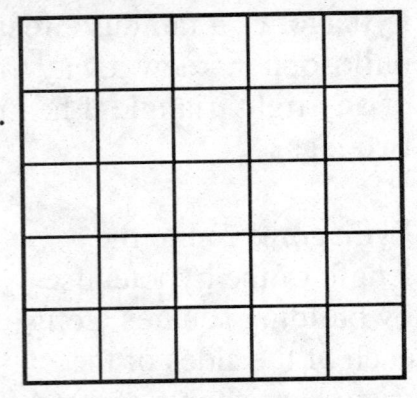

   Figure H

10a. Triangle *GHJ* below is the same triangle you built in the circle on page 124. Use the following steps to calculate the length of *GJ*.

10b. To calculate the area of the square built on Leg *GH*, press [3.3] [x²].

   Area _____ cm²

10c. To calculate the area of the square built along Leg *HJ*, press [3.3] [x²].

   Area _____
   Put in units.

10d. Add the area of the two smaller squares to find the area of the large square.

   Area of _____
   Put in units.

10e. Now take the square root to find the length of Hypotenuse *GJ*.

   Hypotenuse *GJ* ☐.☐6☐☐☐ cm

   Figure J

10f. Hypotenuse *GJ* is the same length as Chord *AB* on page 124. Was your measurement on page 124 close to the calculated answer?

   Answer _____
   Yes or No

Maneuvers with Circles   © David A. Page

Circles, Chords, and the Pythagorean Theorem 127

11. Show that the area of the squares built along the legs is equal to the area of the one large square built along the hypotenuse using the steps in Problems 11 and 12.

   a. Build a square along each leg of the right triangle below.

   b. Shade or color one of the squares.

   c. Cut out the triangle and the squares.

   d. Glue the triangle on a different piece of paper.

   e. Cut and arrange the squares to build a larger square along the hypotenuse of the triangle.

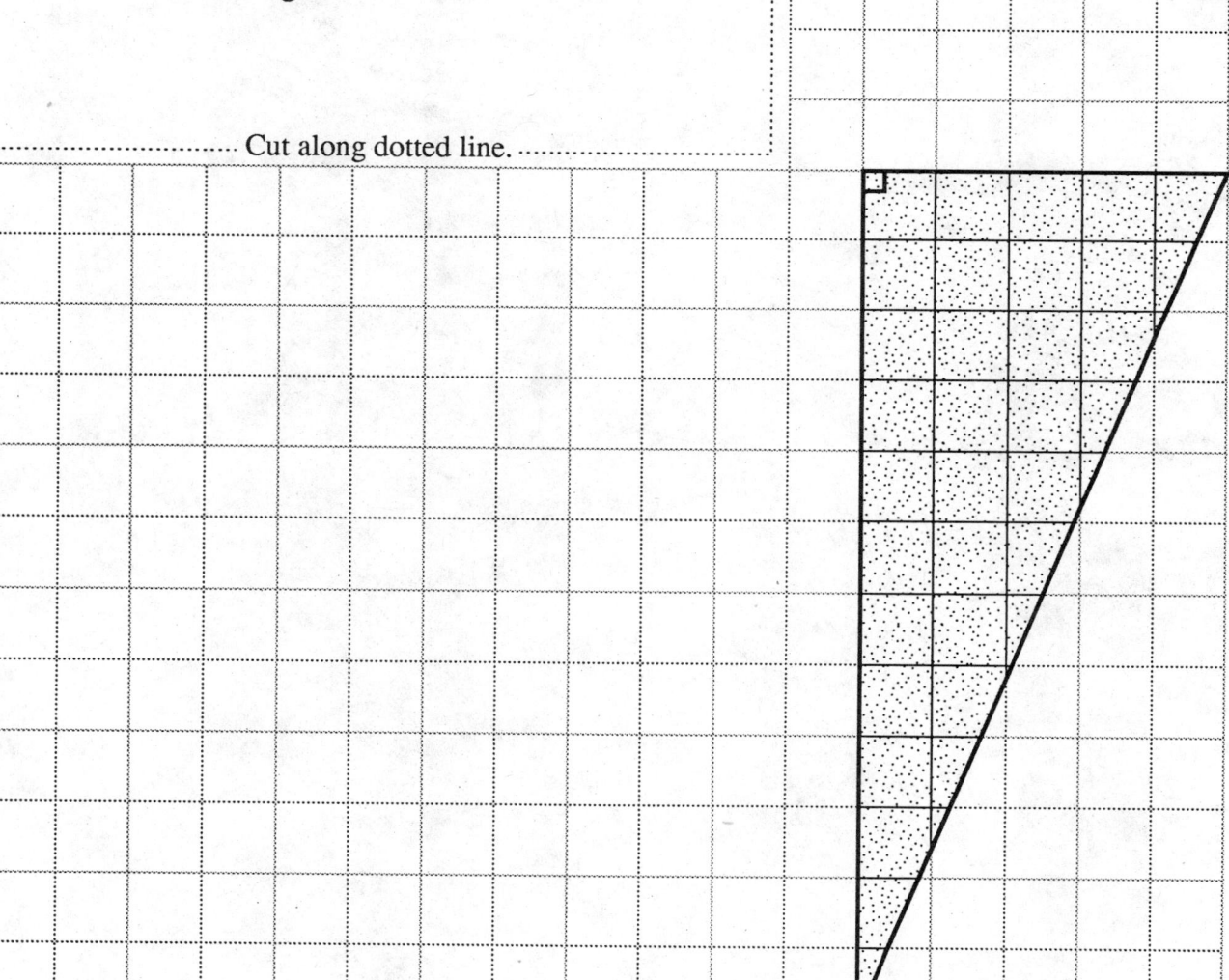

.... Cut along dotted line. ....

Figure K

© David A. Page *Maneuvers with Circles*

Use your figure from Problem 11 to complete the following problems.

12a. Area of square built on one leg _____ cm²

12b. Area of square built on other leg _____
Put in units.

12c. Area of square built on hypotenuse _____
Put in units.

12d. Length of hypotenuse _____ cm

Circles, Chords, and the Pythagorean Theorem   129

13. Measure Chord *KM* to the nearest 0.1 cm.

    Answer _____ cm

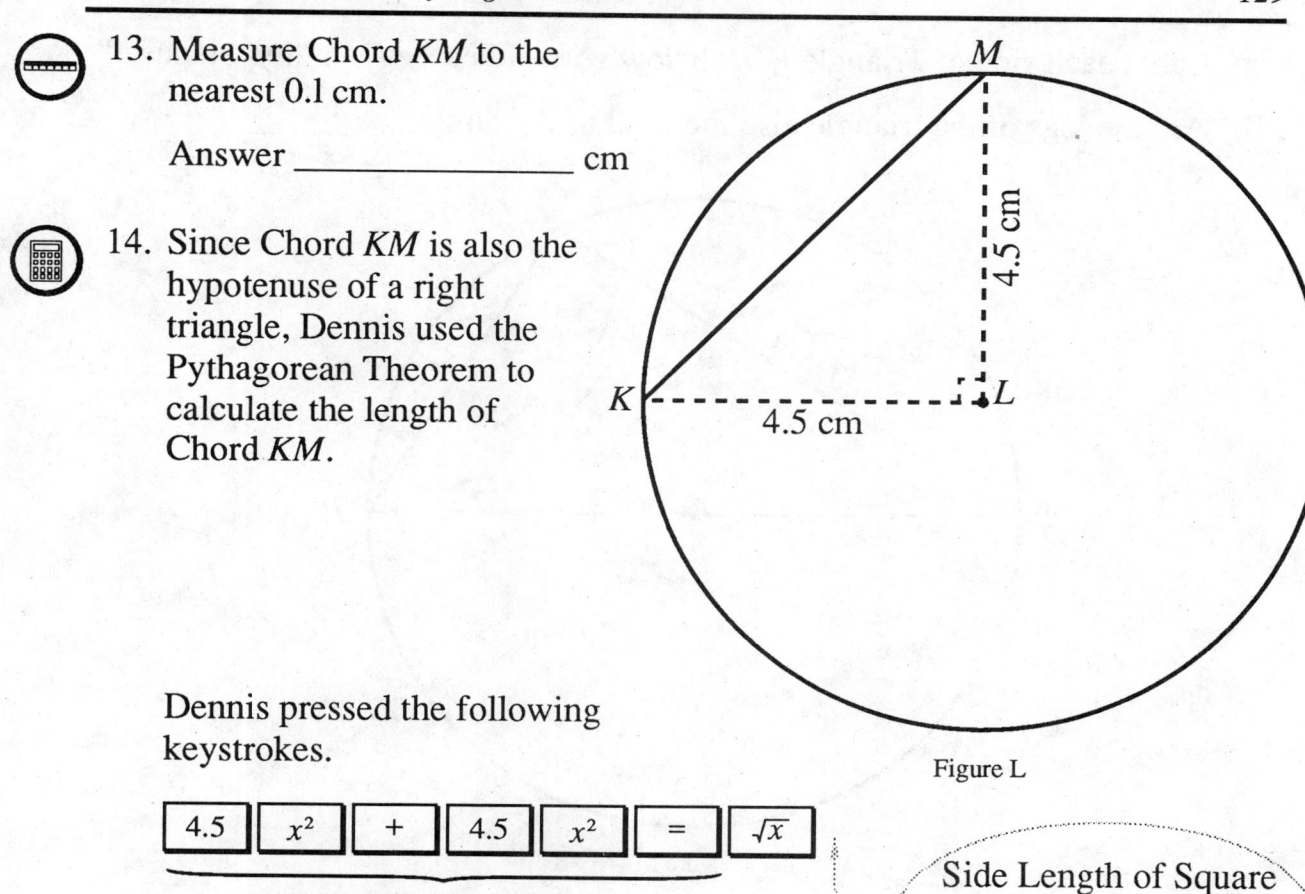

14. Since Chord *KM* is also the hypotenuse of a right triangle, Dennis used the Pythagorean Theorem to calculate the length of Chord *KM*.

Figure L

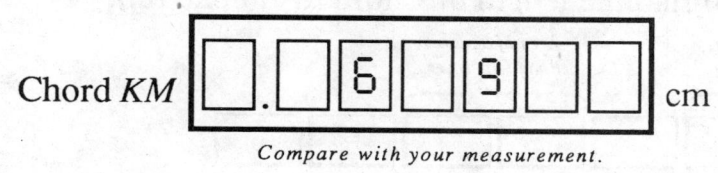

(with labels: M, 4.5 cm, K, 4.5 cm, L)

Dennis pressed the following keystrokes.

| 4.5 | x^2 | + | 4.5 | x^2 | = | $\sqrt{x}$ |

Area of Square Built along Hypotenuse

Side Length of Square
or
Length of Hypotenuse

14a. Put a loop around the keystrokes that give the area of the square built along one of the legs.

14b. Put a box around the keystrokes that give the area of the square built along the other leg.

14c. Use Dennis' keystrokes to calculate the length of Chord *KM*.

Chord *KM* ☐ . ☐ 6 9 ☐ cm

Compare with your measurement.

130   Chapter 9

15a. Label each side of Triangle *RQP* below with either "leg" or "hypotenuse."

15b. Are the legs of the triangle also the radii of the circle? _____
                                                                 Yes or No

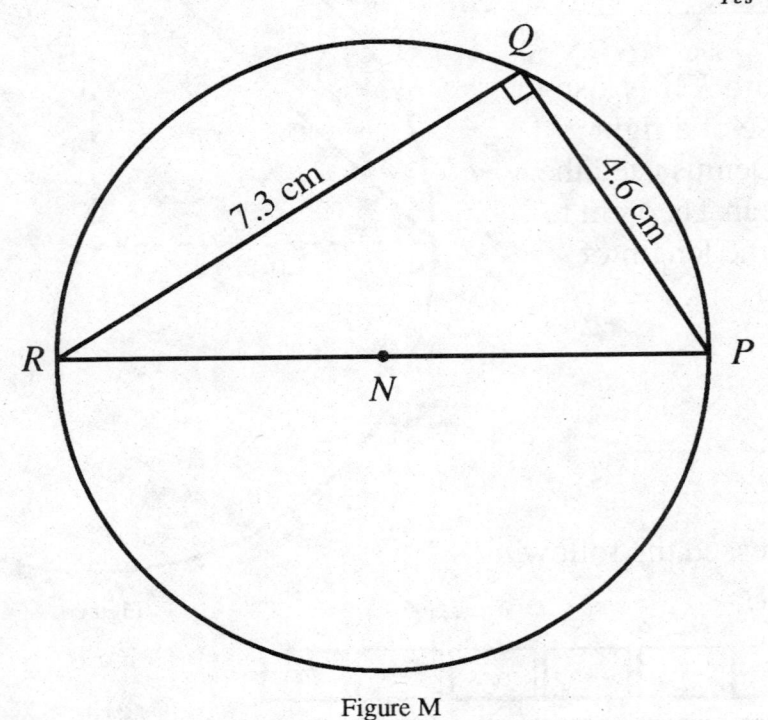

[Figure M: Circle with center N, triangle RQP inscribed. RQ = 7.3 cm, QP = 4.6 cm, right angle at Q. RP is a diameter passing through N.]

Figure M

15c. Hypotenuse *RP* is also a chord of the circle. What is special about this chord?

_____

16. Measure Chord *RP* to the nearest 0.1 cm.

    Measured length _____
                      Put in units.

17a. Fill in the keystrokes to find the length of Chord *RP* in one run.

    Area of Square   Area of Square
    along One Leg   along Other Leg
    ⎴   ⎴
    ☐ ☐ ☐   ☐ ☐   = ☐
    ⎵_____⎵   ⎵___⎵
    Area of Square Built   Side of Square
    along Hypotenuse   or
                              Hypotenuse

17b. Use the Pythagorean Theorem to calculate the length of Chord *RP*.

    Chord *RP* _____ cm
                R to the nearest tenth.

Maneuvers with Circles   © *David A. Page*

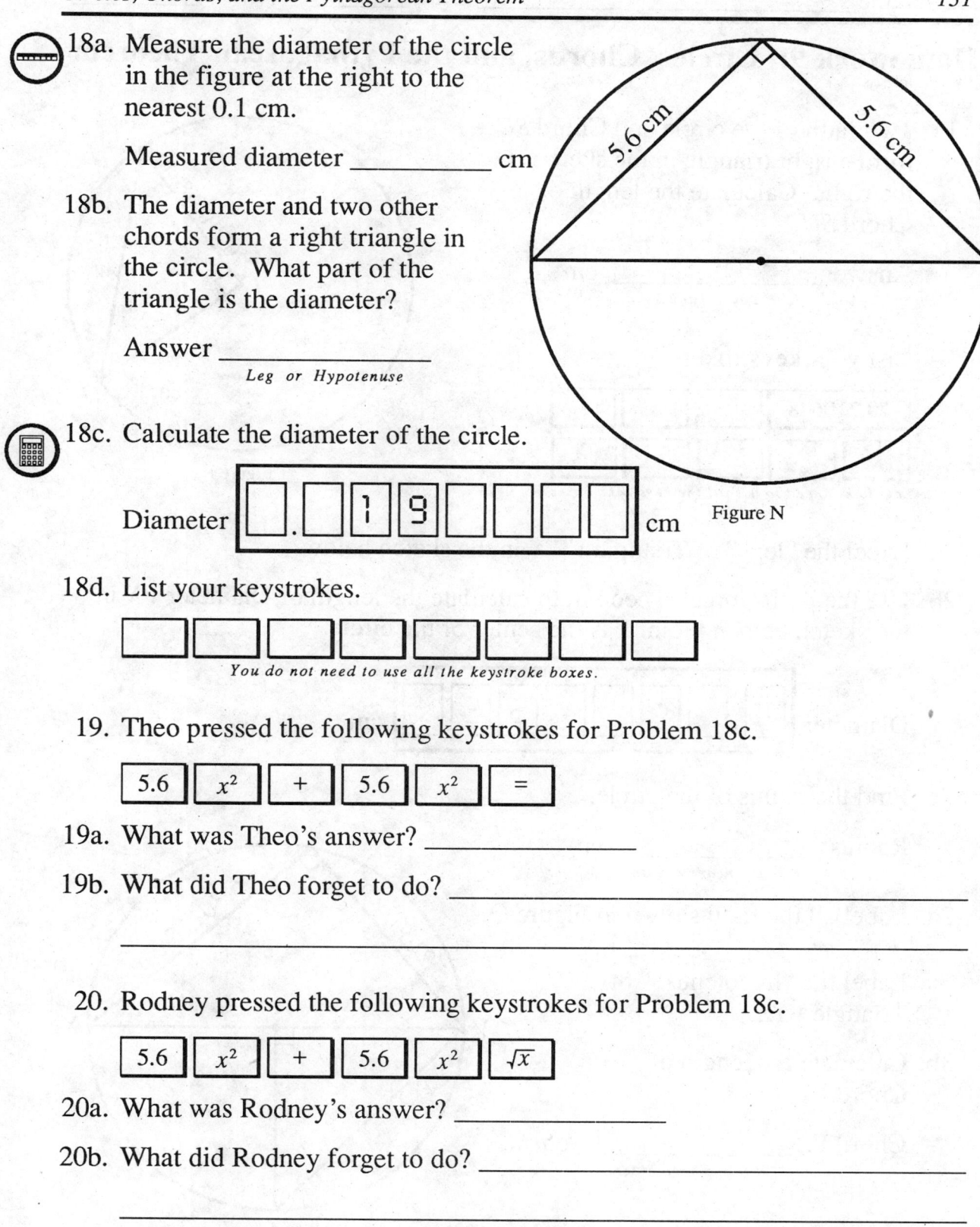

Circles, Chords, and the Pythagorean Theorem   131

18a. Measure the diameter of the circle in the figure at the right to the nearest 0.1 cm.

   Measured diameter _____ cm

18b. The diameter and two other chords form a right triangle in the circle. What part of the triangle is the diameter?

   Answer _____
   Leg or Hypotenuse

18c. Calculate the diameter of the circle.

   Diameter [ ][.][1][9][ ][ ][ ] cm   Figure N

18d. List your keystrokes.

   [ ][ ][ ][ ][ ][ ][ ][ ]
   You do not need to use all the keystroke boxes.

19. Theo pressed the following keystrokes for Problem 18c.

   [5.6][x^2][+][5.6][x^2][=]

19a. What was Theo's answer? _____

19b. What did Theo forget to do? _____
   _____

20. Rodney pressed the following keystrokes for Problem 18c.

   [5.6][x^2][+][5.6][x^2][$\sqrt{x}$]

20a. What was Rodney's answer? _____

20b. What did Rodney forget to do? _____
   _____

© David A. Page   *Maneuvers with Circles*

132 Name _____ Date _____ Class _____

Homework 9: Circles, Chords, and the Pythagorean Theorem

1a. Two radii of the circle and Chord SU form a right triangle in the sketch at the right. Calculate the length of Chord SU.

  Answer _____ cm
       You'll know.

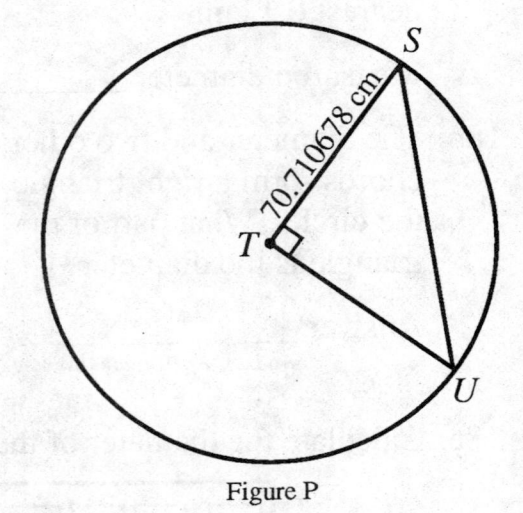

1b. List your keystrokes.

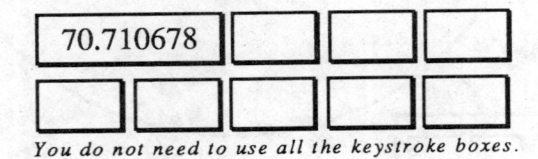

70.710678			

You do not need to use all the keystroke boxes.

Figure P (circle with points S, T, U; T is center with right angle; label 70.710678 cm on radius TS)

2a. Label the "legs" of Triangle VWX in the sketch below.

2b. Use the Pythagorean Theorem to calculate the length of Diameter VX in the sketch below. Point Y is the center of the circle.

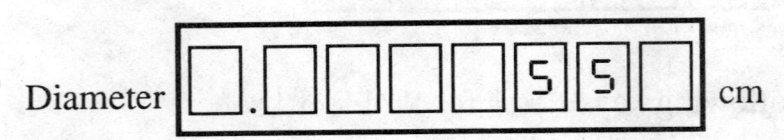

  Diameter ☐.☐☐☐55☐ cm

2c. Find the radius of the circle.

  Radius _____ cm
       R to the nearest tenth.

2d. Label all the radii shown in Figure Q.

3a. Label the "hypotenuse" of Triangle VYZ.

3b. Calculate the length of Chord VZ.

  Chord VZ _____ cm
       R to the nearest tenth.

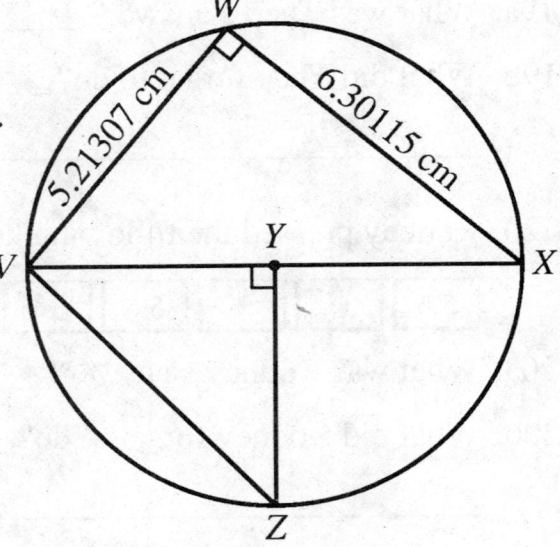

Figure Q (circle with points V, W, X, Y, Z; Y is center; VW = 5.21307 cm, WX = 6.30115 cm; right angles at W and Y)

Maneuvers with Circles   © *David A. Page*

Circles, Chords, and the Pythagorean Theorem   133

4. Triangles *GKJ* and *HKJ* are congruent in the sketch at the right. The length of Chord *GH* is 65 cm. Label the "legs" of the two smaller triangles. Use the Pythagorean Theorem to find the radius of the circle.

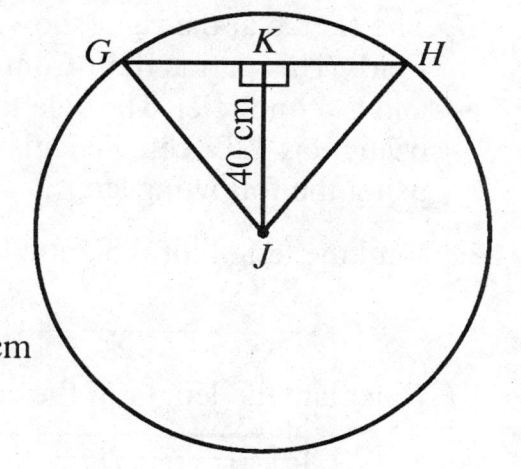

   Radius ☐ ☐ . ☐ ☐ ☐ ☐ cm
         (with 1 and 3 filled in)

Figure R

5a. Four chords build a square in the figure at the right. Measure each chord to the nearest 0.1 cm to find the perimeter.

   Measured perimeter _____
                    Put in units.

5b. Use the radii of the circle to calculate the perimeter of the square. Show your work.

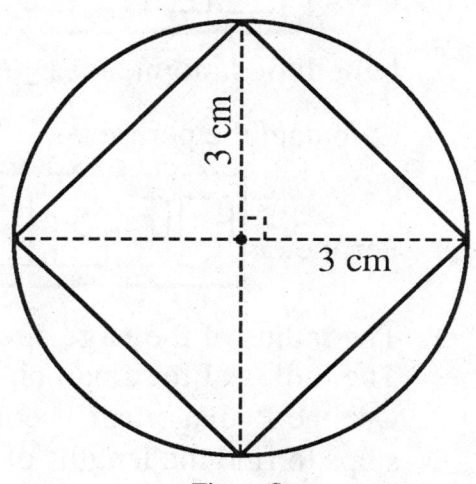

Figure S

   Perimeter _____ cm
   R to the nearest tenth.
   Compare with your measurement.

★ 6. The sketch at the right is built from a square and a circle. The radius of the circle is 78.488853 cm. Calculate the length of the square's diagonal. Show your work.

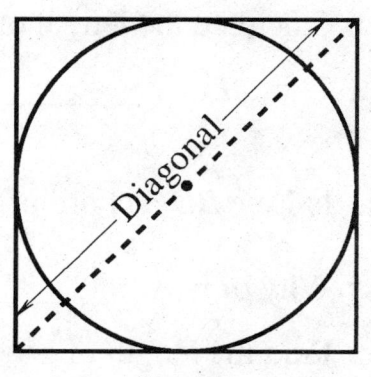

Figure T

   Diagonal _____ cm
        You'll know.

© David A. Page   Maneuvers with Circles

134 Chapter 9

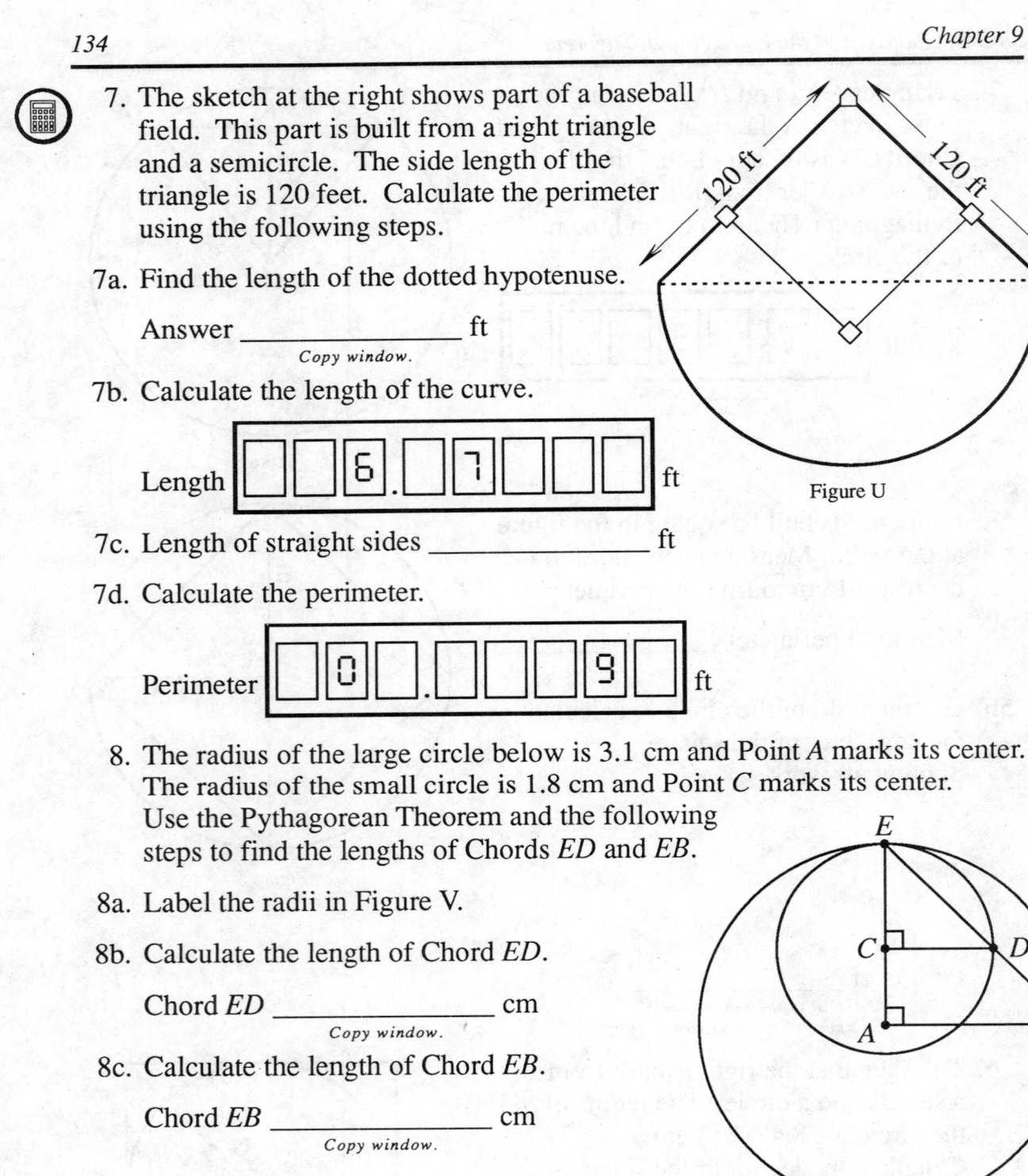

7. The sketch at the right shows part of a baseball field. This part is built from a right triangle and a semicircle. The side length of the triangle is 120 feet. Calculate the perimeter using the following steps.

7a. Find the length of the dotted hypotenuse.

Answer _____ ft
Copy window.

7b. Calculate the length of the curve.

Length ☐ ☐ 6 . ☐ 7 ☐ ☐ ft

7c. Length of straight sides _____ ft

7d. Calculate the perimeter.

Perimeter ☐ 0 ☐ . ☐ ☐ 9 ☐ ft

Figure U

8. The radius of the large circle below is 3.1 cm and Point *A* marks its center. The radius of the small circle is 1.8 cm and Point *C* marks its center. Use the Pythagorean Theorem and the following steps to find the lengths of Chords *ED* and *EB*.

8a. Label the radii in Figure V.

8b. Calculate the length of Chord *ED*.

Chord *ED* _____ cm
Copy window.

8c. Calculate the length of Chord *EB*.

Chord *EB* _____ cm
Copy window.

9a. Is Line *BD* a chord of one of the circles? _____
Yes or No

Why or why not? _____

9b. Find the length of Line *BD*.

Answer _____ cm
Measure to check.

Figure V

Maneuvers with Circles © David A. Page

Name _____ Date _____ Class _____ 135

10. Grazing Goats

A farmer tied his goat, Billy, to a stake so that the goat could graze. A long rope or *tether* was used to tie the goat to the stake. Two days later, the farmer noticed Billy had eaten all the grass within a circular region.

1. The sketch at the right shows the circular region of grass that Billy ate. Notice Billy's tether is 15 feet long. What is the area of this circular region in square feet (ft^2)?

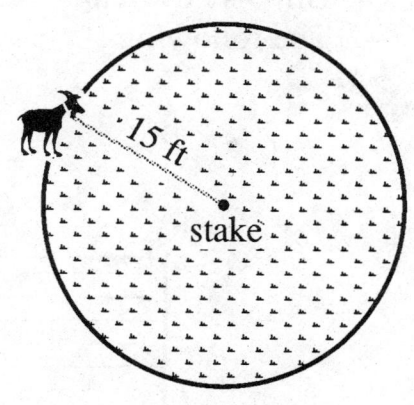

   Area ☐☐☐.8☐8☐☐ ft^2

   Figure A

2. Since Billy ate all the grass in this circular region, the farmer moved him to a place with fresh grass. In the sketch at the right, Billy is tied to a fence post with the same tether. Billy can only graze on one side of the fence. His grazing region is a semicircle. What is the area of this semicircular region?

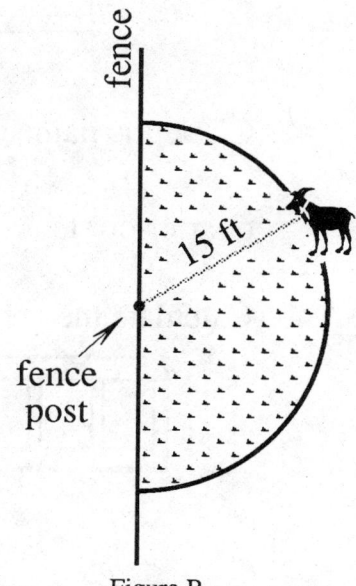

   Area _____ ft^2
   Copy window.

   Figure B

© David A. Page

Maneuvers with Circles

136 Chapter 10

3. The next day, the farmer moved Billy into a yard enclosed by a rectangular fence. He tied Billy to a stake in the corner of the yard (so he wouldn't eat the petunias).

3a. What part of a circle can Billy graze?

Answer _____

3b. Calculate the area of the grass Billy can graze.

Area _____ ft²
Copy window.

3c. Compare your answers to Problems 1, 2, and 3b.

Figure C

4. Billy is now tethered to the corner of a small rectangular barn. The following sketch shows the region that Billy can graze.

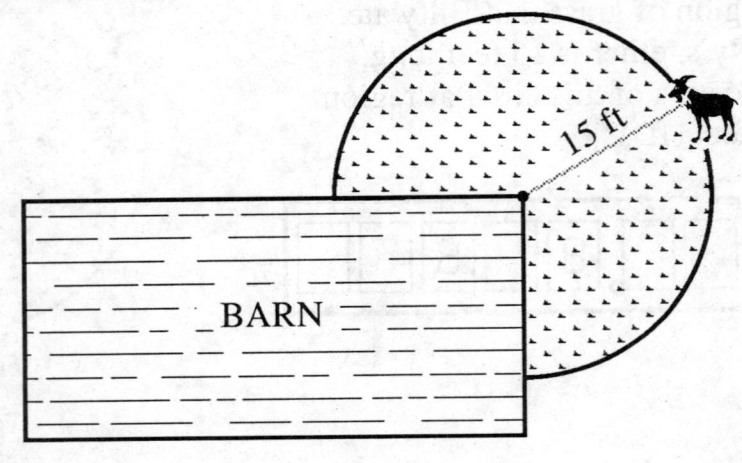

Figure D

4a. Circle the name of the largest region Billy can graze.

quarter-circle   semicircle $\frac{3}{4}$ circle   full circle

4b. Calculate the area that Billy can graze.

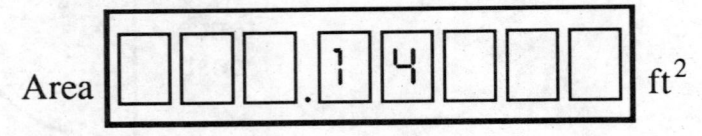

Area ☐☐☐.☐☐☐ ft² (with 1 and 4 shown)

Maneuvers with Circles © *David A. Page*

Grazing Goats

5. Billy is attached to a stake at the corner of a barn in Figure E.

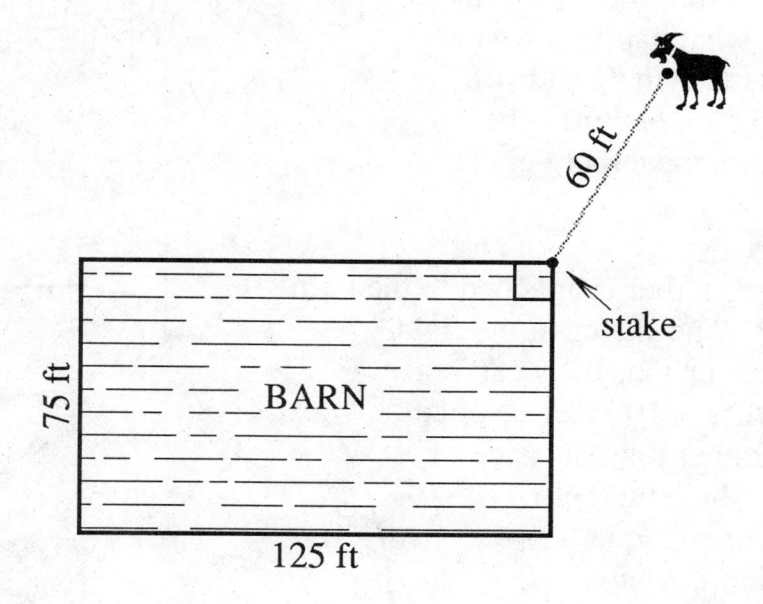

   [Figure showing a barn 125 ft × 75 ft with a stake at the upper-right corner and a 60 ft tether extending to a goat.]

   Figure E

   When the tether is tight, it acts like a large compass. Use your compass and the following steps to draw a picture of the area where Billy can graze.

   a. Place the sharp point of your compass on the stake in Figure E.

   b. Place the pencil on the end of the tether.

   c. Draw the part of the circle where Billy can graze. (Beware! The tether cannot go inside the barn.)

   d. Compare your drawing with the picture at the right.

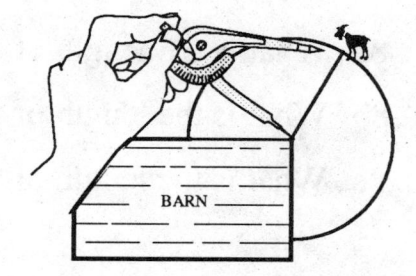

6. Shade the region Billy can graze in Figure E.

7. Calculate the area that Billy can graze. Show your work.

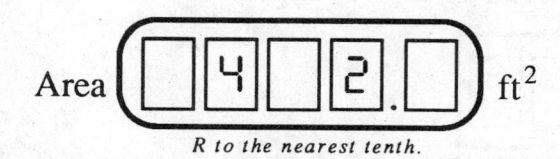

   Area ⬜ 4 ⬜ 2 . ⬜ ft²

   R to the nearest tenth.

© David A. Page   *Maneuvers with Circles*

138   Chapter 10

So far, the tether has been shorter than
the sides of the barn. The picture at
the right shows what happens when
the tether is longer than the side of
the barn. When the goat turns the
corner of the barn, the tether bends.

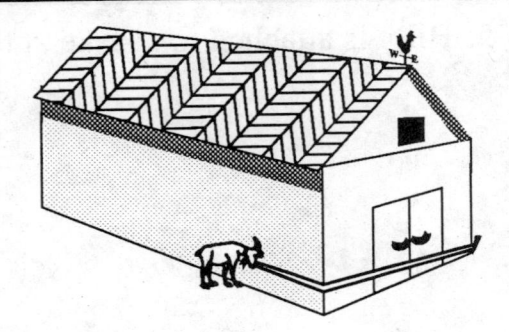

8. Now Billy's tether is attached to the barn
at Point B. The sketch at the right
shows the area that Billy can graze.
Billy's tether is 100 feet. Notice
that it is longer than one side of
the barn. The length of CD
is the radius of a quarter-
circle. Find the length
of CD using the steps
below. Label CD on
the figure.

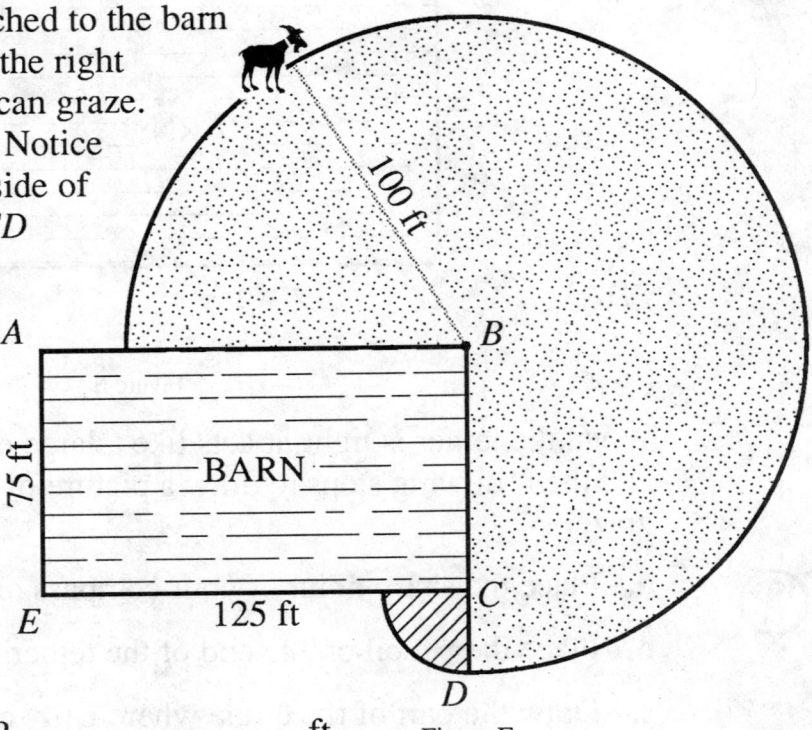

75 ft

BARN

E   125 ft   C
                     D

Figure F

8a. What is the length of BD? _____ ft

8b. What is the length of BC? _____ ft

8c. What is the length of CD? _____ ft
    Don't forget to show this length on the figure.

9a. Calculate the area of the ░░░░░ region. _____ ft²
    Copy window.

9b. Calculate the area of the ▨▨▨▨ region. _____ ft²
    Copy window.

9c. Add the two areas to find the total grazing area in Figure F.

Area [ ][4][ ][2].[ ][ ] ft²

Maneuvers with Circles   © David A. Page

Grazing Goats 139

10. Now Billy is on an even longer tether. The sketch at the right shows the area that Billy can graze. Billy's tether is 150 feet. Notice that it is longer than both sides of the barn.

The lengths of *DE* and *AF* are the radii of the quarter-circles. Find *DE* and *AF* using the following steps. Write these lengths on the figure.

Figure G

10a. What is the length of *BE*? _____ ft

10b. What is the length of *DE*? _____ ft

10c. What is the length of *BF*? _____ ft

10d. What is the length of *AF*? _____ ft

11a. Calculate the area of the ⋯ region. _____ ft^2
Copy window.

11b. Calculate the area of the ⋯ region. _____ ft^2
Copy window.

11c. Calculate the area of the ⋯ region. _____ ft^2
Copy window.

11d. Add the three areas to find the total grazing area in Figure G.

Area ☐ 7 ☐ ☐ ☐ . 7 ☐ ☐ ft^2

© David A. Page   *Maneuvers with Circles*

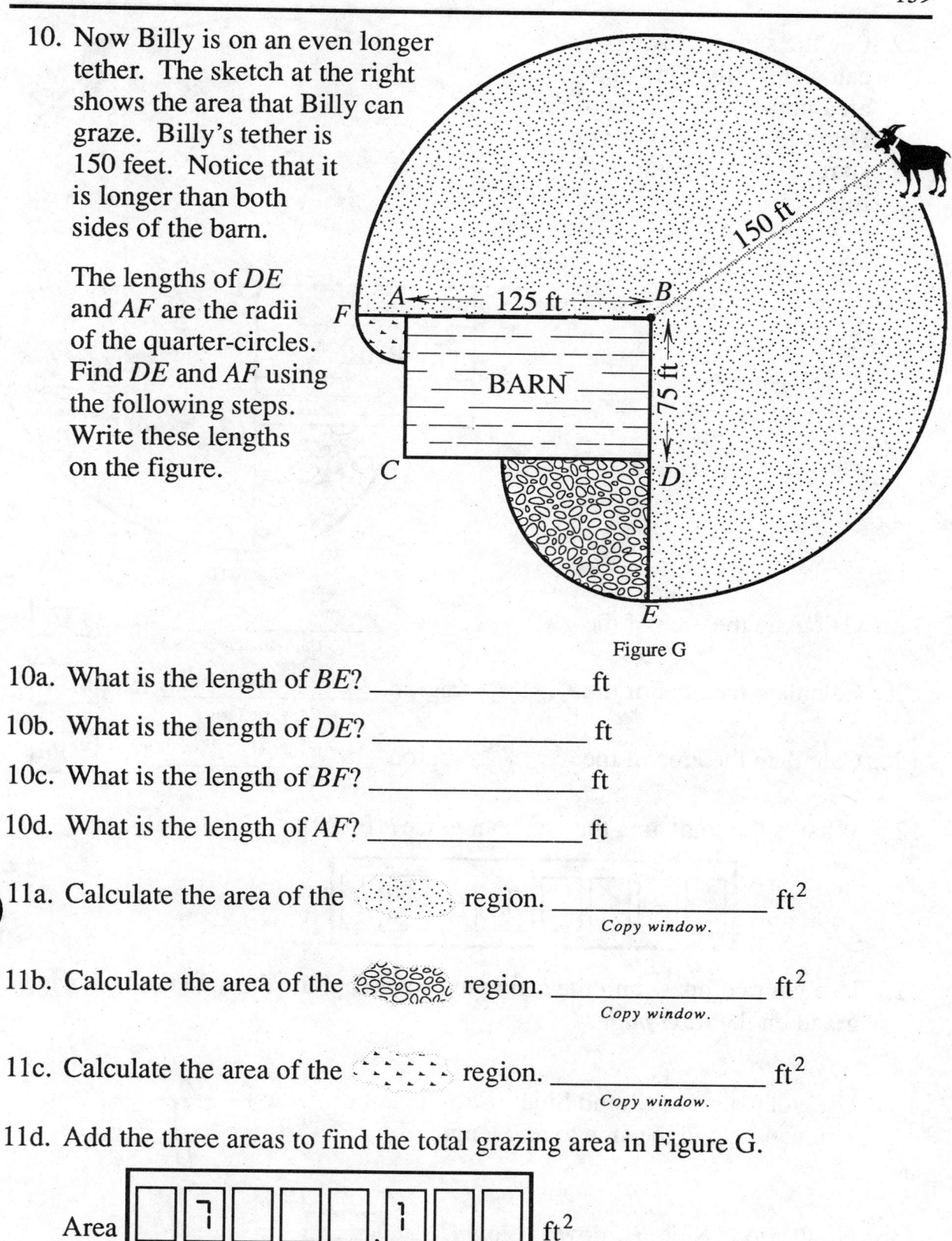

140   Chapter 10

12. Use the steps below to calculate Billy's grazing area in the sketch at the right. Don't press [AC/ON] after you start or you will erase the memory.

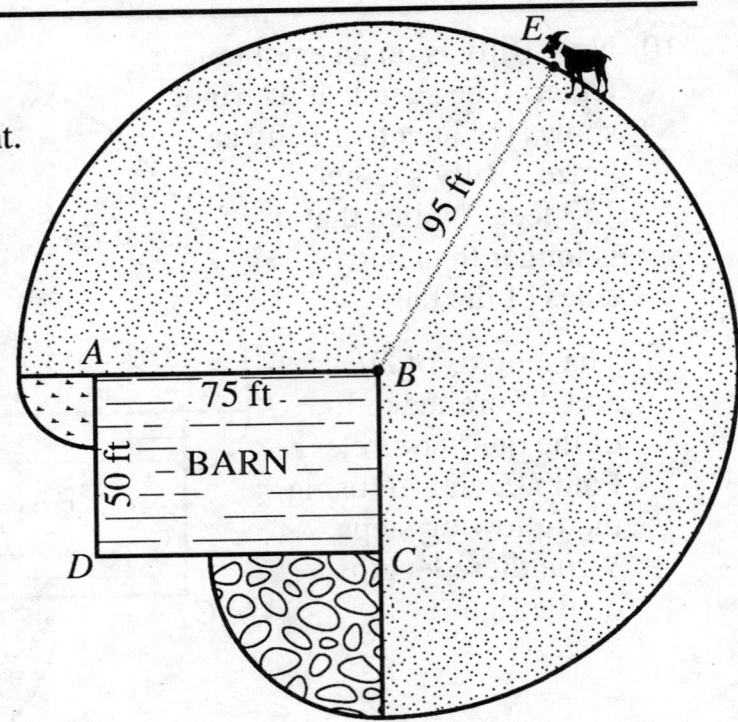

Figure H

12a. Calculate the area of the ░░░ region. _____ ft² [STO]
     Copy window.

12b. Calculate the area of the ░░░ region. _____ ft² [SUM]
     Copy window.

12c. Calculate the area of the ○○○ region. _____ ft² [SUM]
     Copy window.

12d. What is the total area the goat can graze? Press [RCL].

Total area ☐ ☐ ☐ ☐ 9 . 2 ☐ ☐ ft²

13. Use your compass and the following steps to draw the area the goat can graze on the **next page**.

13a. Use your ruler to extend Side AB to the left of A. You can extend the side to any length.

13b. Now extend Side BC down below C.

Maneuvers with Circles © David A. Page

Grazing Goats 141

Figure J

13c. Put the compass point on *B* and the compass pencil on *E*. Draw three-fourths of a circle with a radius of 90 feet.

13d. Put the compass point on *C*. Move the compass pencil to make the new radius on the goat's path directly below *C*. Draw a quarter-circle.

13e. How long is the radius of this quarter-circle? _____ ft

13f. Put the compass point on *A* and move the compass pencil to the new radius on the goat's path. Draw the small quarter-circle.

13g. How long is the radius of this small quarter-circle? _____ ft

13h. Shade each of the three regions with a different color or design.

14. Calculate the area of the three shaded regions. Use your calculator's memory to find the total area.

14a. Area of three-quarter circle _____ ft^2
Copy window.

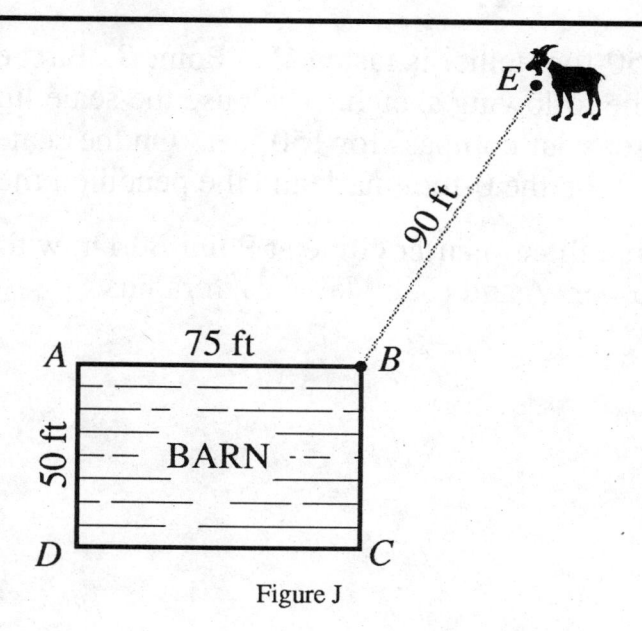

14b. Area of larger quarter-circle _____ ft^2
Copy window.

14c. Area of smaller quarter-circle _____ ft^2
Copy window.

14d. Total area ☐ ☐ ☐ 1 8 . ☐ ☐ ☐ ft^2

© *David A. Page*   *Maneuvers with Circles*

142　　　　　　　　　　　　　　　　　　　　　　　　　　　　　　　　Chapter 10

15a. Imagine a 150-foot tether is fastened at Point B. First extend the sides of the barn in the following sketch. Then use the scale line at the bottom of the page to set your compass for 150 feet. On the scale line, put the compass point on the 0-foot mark and the pencil on the 150-foot mark.

15b. Draw the large three-quarter circle at Point B. Draw the two quarter-circles at Corners A and C and label each radius.

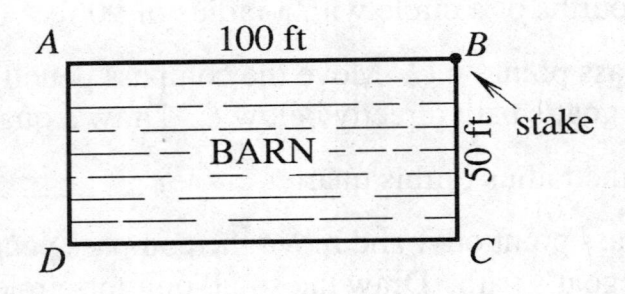

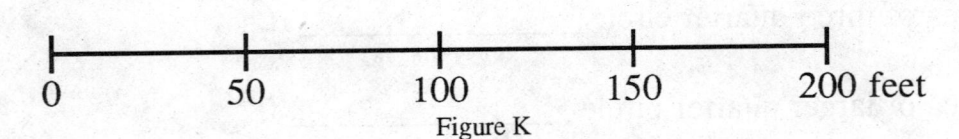

```
A        100 ft         B
 ┌─────────────────────┐  • stake
 │ ─ ─ ─ ─ ─ ─ ─ ─ ─ ─ │
 │ ─ ─ BARN ─ ─ ─ ─ ─ │ 50 ft
 │ ─ ─ ─ ─ ─ ─ ─ ─ ─ ─ │
 └─────────────────────┘
D                         C
```

|―――――|―――――|―――――|―――――|
0   50   100   150   200 feet

Figure K

15c. Use your calculator's memory to find the grazing area. Label the picture with the area of each piece.

15d. What is the total grazing area? ☐☐8☐☐.8☐☐ ft²

Maneuvers with Circles　　　　　　　　　　　　　　　　　　　　　© *David A. Page*

Grazing Goats

★16a. Now there is a 50-foot fence connected to the barn.
The tether is still 150 feet and is fastened at Point *B*. Extend *AB* and *CE*.
Draw the region that Billy can graze at Point *B*, Point *A*, and Point *E*.
Hint: One of the grazing areas is a semicircle.

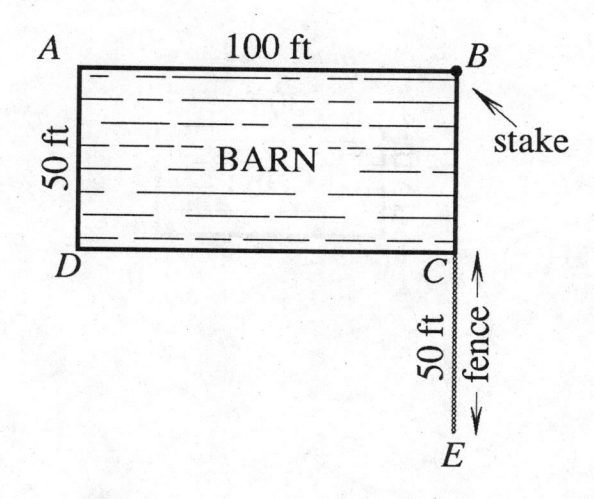

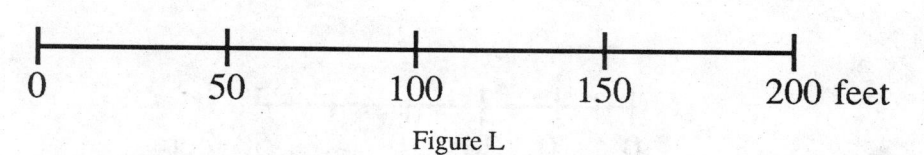

Figure L

16b. Do you think Billy will reach more or less grass than in Problem 15?

Answer _____
      More or Less

Tell why you think so. _____

_____

16c. Calculate the total grazing area. Use the calculator's memory.

Total area (8 _ _ _ . _) ft²

R to the nearest tenth.
Compare with Problem 15d.

© David A. Page   *Maneuvers with Circles*

17. The following picture shows a field at the Guzzling Goat Estate. An old, abandoned log cabin sits in the middle of the field. Each dot on the map represents an oak tree. A fence protects the garden from the goat.

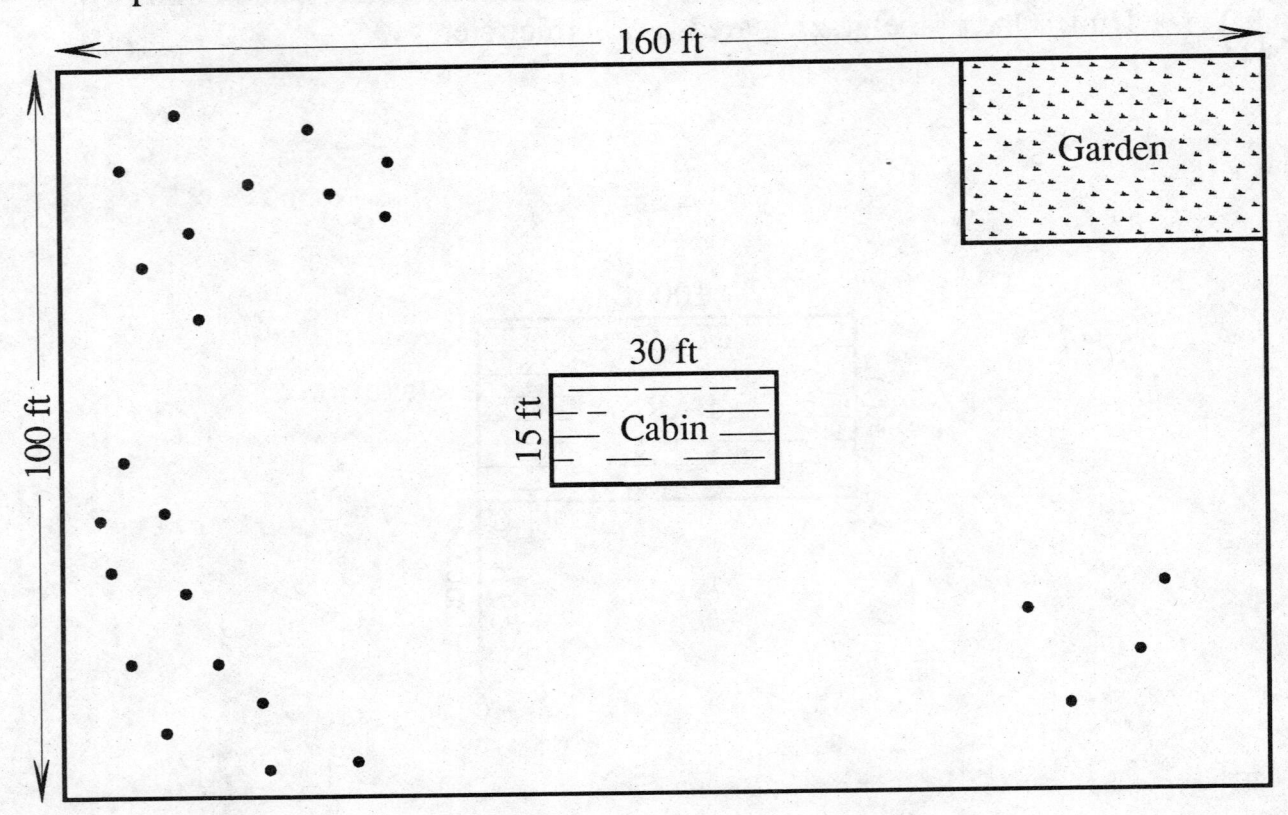

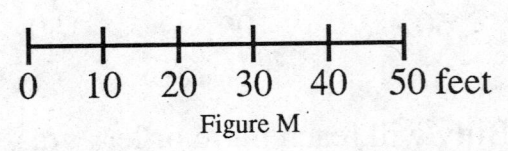

Figure M

17a. A farmer wants to attach his goat to a corner of the cabin with a 40-foot rope. Decide which corner gives the goat the most grazing area. Draw a dot at that corner.

17b. Draw the region the goat can graze.

17c. How much area can the goat graze? Show your work on the sketch.

Area _____ ft^2

Copy window.

Grazing Goats   145

18a. The square barn in the following sketch has a side length of 100 feet. A goat is fastened to a stake with a 200-foot tether. Draw the region that the goat can graze and label the radius of each region. Extend all the sides of the barn.

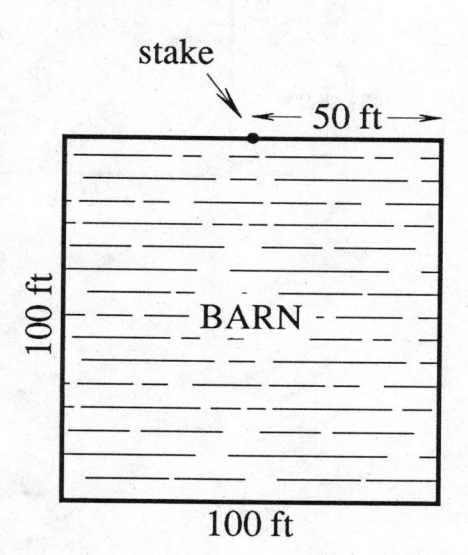

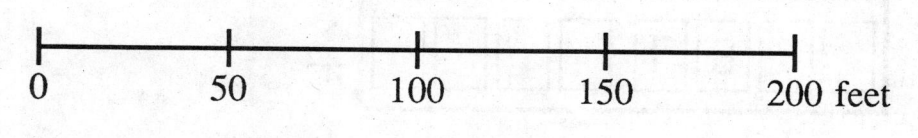

stake
← 50 ft →

100 ft

BARN

100 ft

├──────┼──────┼──────┼──────┤
0   50   100   150   200 feet

Figure N

18b. Calculate the area that the goat can graze.

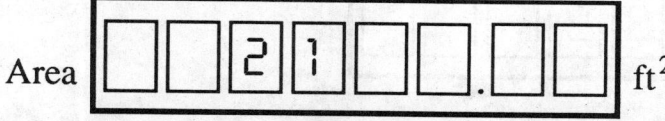

Area ☐ ☐ 2 ☐ 1 ☐ ☐ . ☐ ☐ ft²

© David A. Page   *Maneuvers with Circles*

146   Chapter 10

19a. A 50-foot fence and a 40-foot fence meet at a right angle. A goat is fastened to a stake where the fences meet. If the goat's tether is 60 feet long, what is the area of grass he can reach? Draw the area.

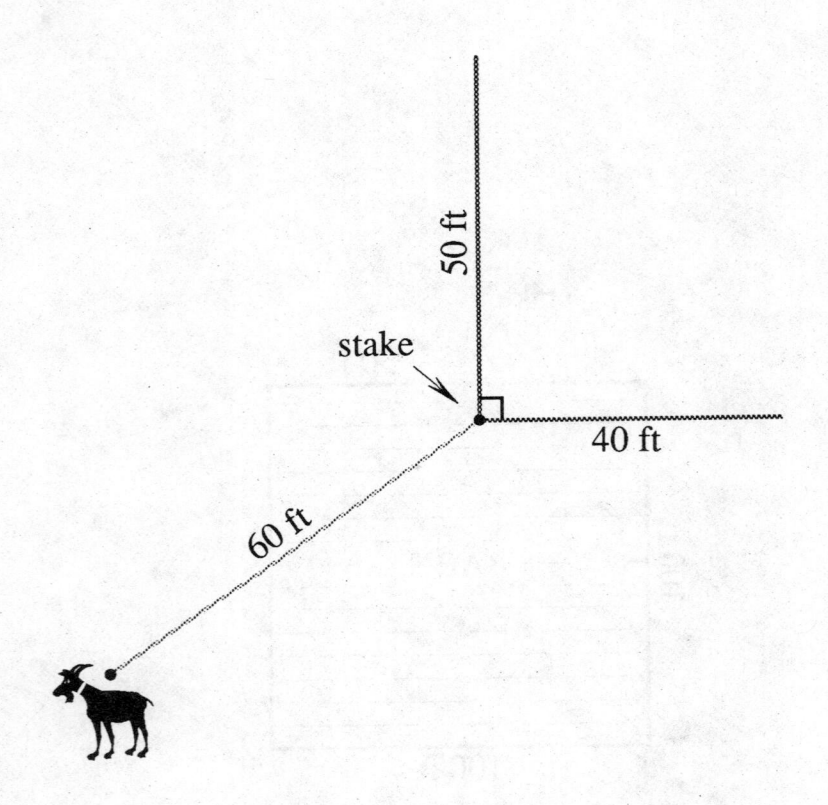

Figure P

19b. Calculate the area of the region the goat grazes.

Area [ ][ ][6][7].[ ][ ][ ] ft²

★ 20. Calculate the perimeter of the region the goat grazes.

Perimeter [ ][ ][ ].[9][9][ ][ ] ft

Grazing Goats   147

21a. A 50-foot fence and a 40-foot fence meet at a right angle. A goat is fastened where the fences meet. If the goat's tether is 61.890377 feet long, what is the area of grass he can reach? Draw the area.

stake

50 ft

40 ft

61.890377 ft

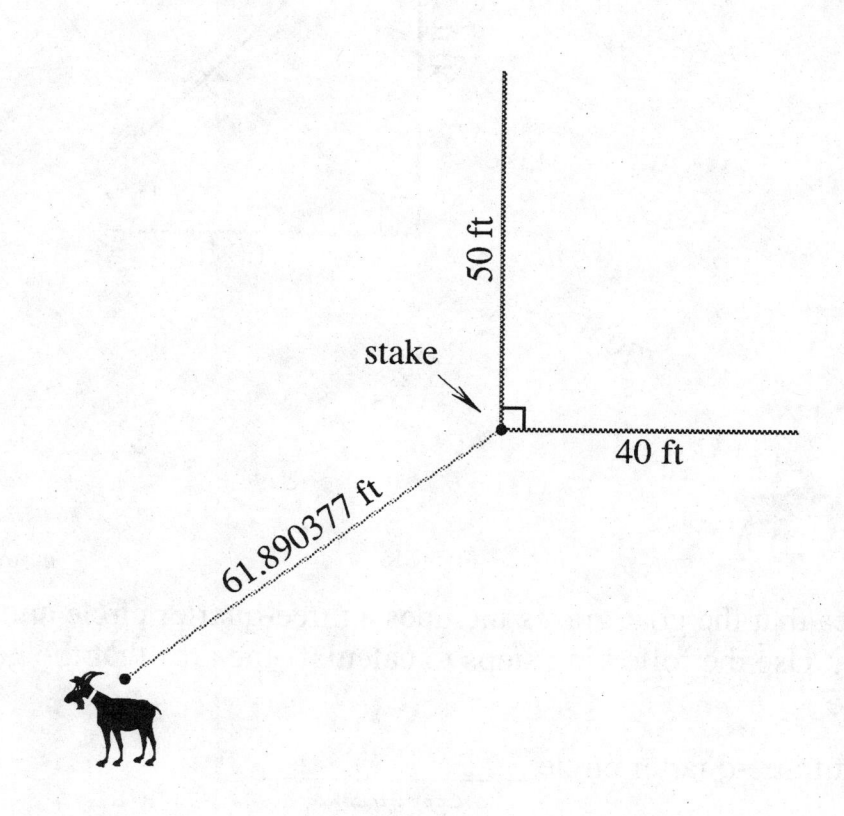

Figure Q

21b. Calculate the area. Show your work.

Area _____ ft^2
      You'll know.

© *David A. Page*   *Maneuvers with Circles*

148 Chapter 10

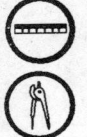

22. A goat is tethered to a triangular-shaped fence in the following sketch. Draw and shade the region that the goat is able to graze. Remember to extend the sides of the fence when necessary.

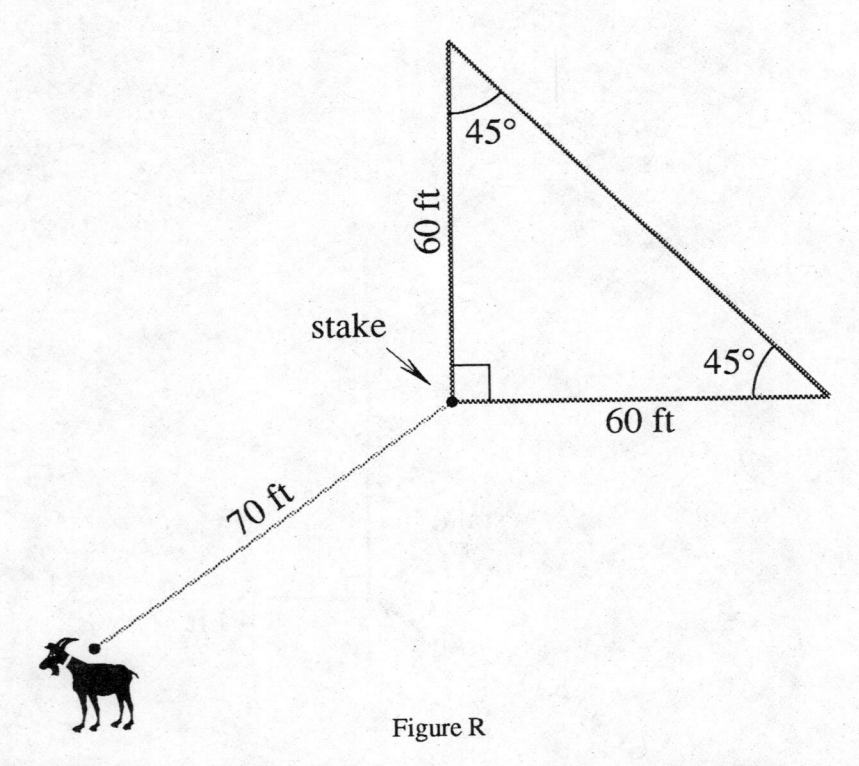

Figure R

23. The area that the goat grazes includes a three-quarter circle and two small sectors. Use the following steps to calculate the area that the goat is able to graze.

23a. Area of three-quarter circle _____ ft²
     Copy window.

23b. Radius of small sector _____ ft

23c. The central angle of the small sector and the 45° angle form a straight line. Find the degree measure of the small sector.

   Degree measure _____°

23d. Area of one small sector _____ ft²
     Copy window.

23e. Grazing area ☐☐☐8☐.☐7☐ ft²

Maneuvers with Circles © *David A. Page*